BONDING AND STRUCTURE OF SOLIDS

BONDING AND STRUCTURE OF SOLIDS

PROCEEDINGS OF
A ROYAL SOCIETY DISCUSSION MEETING
HELD ON 20 AND 21 SEPTEMBER 1990

ORGANIZED AND EDITED BY
R. HAYDOCK, J. E. INGLESFIELD
AND J. B. PENDRY

LONDON
THE ROYAL SOCIETY
1991

Printed in Great Britain for the Royal Society
by the
University Press, Cambridge

ISBN 0 85403 435 8

First published in *Philosophical Transactions of the Royal Society of London*, series A, volume 334 (no. 1635), pages 391–556

The text paper used in this publication is alkaline sized with a coating which is predominantly calcium carbonate. The resultant surface pH is in excess of 7.5, which gives maximum practical permanence.

British Library Cataloguing in Publication Data

Bonding and structure of solids: proceedings of a Royal Society discussion meeting held on 20 and 21 September 1990.
1. Solids. Structure
I. Haydock, R. II. Inglesfield, J. E. III. Pendry, J. B. (John Brian) IV. Royal Society
530.41

ISBN 0-85403-435-8

Published by the Royal Society
6 Carlton House Terrace, London SW1Y 5AG

PREFACE

The link between our understanding of electrons in a solid and its structure has come to the fore of condensed matter studies in the past few years. It is no less than the key to our understanding of the solid state. Our two-day Discussion Meeting, held in September 1990, was intended to make a broad review of theoretical developments in this field with the help of leading exponents from the United Kingdom and from abroad.

The papers divide under four headings: forces and structure, bonding aspects of high T_c superconductivity, exchange and correlation, and electrons in reduced symmetry.

Under the first of these headings speakers reviewed progress in our understanding of why certain structures form and to what extent we can extract from some of the very complex quantum theories some simple concepts that can be applied flexibly to guide us in our understanding of materials at a practical level. Structure is related to the underlying forces in a subtly balanced fashion and much work has gone into this field in the past decade.

High T_c superconductivity is a field pregnant with promise of practical application, but also the subject of great argument among theorists. The materials which exhibit the phenomenon have in common a layered structure, which is thought to be central to their electrical properties. Three of the current protagonists give their views, the conclusion from which must be that, although we are far from hearing the final word on the theory, the advent of this whole field has provided great theoretical stimulation.

Exchange and correlation among electrons controls the bonding in solids, and we must understand these phenomena before we can understand the structure of solids. And if we wish to make quantitative calculations we must find some flexible scheme which allows us to apply our theory numerically. Three of the leading experts in the field give us their views.

Finally the section on electrons in reduced symmetry reflects the growing interest of condensed matter theorists in non-periodic structures, such as we encounter in surfaces, in alloys, or in structures too complex for their periodicity to be relevant.

February 1991

R. Haydock
J. B. Pendry
J. E. Inglesfield

CONTENTS

Many-atom interactions in solids

By Volker Heine, I. J. Robertson and M. C. Payne
Cavendish Laboratory, Madingley Road, Cambridge CB3 0HE, U.K.

Computer simulation of complex processes in condensed matter comprises a large and broad research effort. These require good models of the interatomic interactions, valid over a wide range of circumstances. In most processes of interest, the crucial atoms are in positions far from standard bonding patterns, at least temporarily: at surfaces and defects, in clusters and in open structures like silicates, the coordination number varies widely. The difficulty of modelling interatomic interactions in such circumstances arises from the existence of strong many-atom forces, originating from the uncertainty principle and the variational principle of quantum mechanics. Some theory of many-atom interactions, and some evidence for them, will be reviewed briefly. In particular a series based on two-, three-, four-atom, etc., interactions is almost certainly not convergent in some cases. In recent years several empirical and semi-empirical, broadly similar approaches to modelling many-atom interactions have come into use, though there are few hard tests of how good they are. An alternative approach, requiring the largest computations possible, involves the Schrödinger equation for the whole simulation to a high relative accuracy.

1. Introduction

An understanding of bonding in condensed matter has become a matter of major interest, as witness the holding of this conference. By condensed matter we mean the usual range of ordered and disordered crystalline solids with their defects and surfaces, as well as melts, glasses and amorphous materials, and even cluster molecules. The understanding of complex processes in condensed matter is central to a broad range of physics, solid state chemistry, metallurgy and to a growing extent in ceramics and mineralogy, i.e. to much of materials science in its widest sense. For example one would like to know the nature of grain boundaries in group IV semiconductors: do they retain basically a four-fold coordination as in the perfect crystal, or a more metallic structure with higher coordination as in the melt? In metals the segregation of impurities in grain boundaries can weaken the material, even impurities such as arsenic and sulphur which form very stable compounds with metallic elements and hence might be expected to strengthen the material by glueing the grains together: Why? One of the earliest bits of science one learns at school is that metals are ductile as well as conducting heat and electricity, whereas non-metals are brittle. When one asks friends in materials science about this, one gets answers in terms of slip systems for dislocations, etc., but these answers are not relatable at the present time to the basic difference that the bonding electrons in metals are much more mobile and spread out over many atoms compared with non-metals. These examples, let alone the whole subject of surface chemistry, show how little detailed quantitative understanding we have of bonding in condensed matter.

The matter of bonding has come to the fore as computer simulation has developed over the past 25 years. The point here is that a computer simulation is only as good as the model of the interatomic binding that goes into it, and that is still its Achilles heel, as it has been throughout the evolution of the subject.

It must be admitted that we are asking for a lot from interatomic potential. The foremost difficulty is that complex processes of interest usually involve atoms arranged in configurations which are very far from standard bonding geometries. Consider for example the rebonding of the broken bond at a vacancy in silicon: or an interstitial at the top of the potential barrier in diffusion. Our model of the bonding has to be correct for those geometries. These are very different situations from lattice vibrations where the bonding topology remains that of the standard coordination with only moderate deviations from equilibrium.

For the latter the ionic models and bond charge models have been very successful; but even here often only reliable within the harmonic approximation, not for third- and fourth-order anharmonicity. The second requirement is a model of the interatomic bonding applicable over a wide range of coordination number C. Consider two A atoms, together and separated, adsorbed on the surface of a different material B. We need the A–A bonding for $C = 1$ in the presence of B, and if we want to compare it with the cohesive energy of the bulk A solid the same model needs to be applicable for the much larger value $C = 4$ to 12 of the bulk A. The same point is true for studying phase transitions under pressure where materials tend to collapse to denser packing (larger C), e.g. silicates at depth in the Earth's mantle. The last example serves to illustrate a third requirement: the potential model needs to be accurate. The mantle convection sustaining geological plate tectonics is driven by density changes probably caused by phase changes as much as by thermal expansion. To make useful predictive statements about the pressures for such phase changes, the potential model has to be quite accurate in the comparison of what may be quite different structures. It all amounts to a tall order!

At one time it was hoped that the total cohesive energy of the system could be represented to considerable accuracy by a sum of pairwise interactions, supplemented by some three-atom terms for the bond bending forces but not much more. True, an atom has a constant size to a remarkable degree whatever its environment, which indicates that the short range, hard core, repulsive forces are largely pairwise in nature; but it tells us nothing about the softer cohesive interactions at larger distances.

It is not widely recognized that we have to think about many-atom bonding in condensed matter, and 1989 saw the first international conference specifically devoted to that issue (Nieminen *et al.* 1990). In metals this has been realized at the qualitative level for a long time because of the electron-gas nature of metallic electrons, as evidenced by the Cauchy anomaly among the elastic constants. But Mott & Jones (1936) in their pioneering monograph on theoretical solid state physics also treated diamond as a good example of a nearly free-electron gas, and we now know they were right (in a slightly more sophisticated sense than they realized (Heine & Jones 1969)). This is an awful warning not to be too cavalier in thinking about even the best saturated covalent materials. (Of course materials with unsaturated conjugated bonds as in benzene are similar to metals.) The issue is really whether one can write the formation energy U as a rapidly converging sum of pairwise, three-atom, four-atom, etc., interactions

$$U = I_2 + I_3 + I_4 + \dots . \quad (1.1)$$

Or must one think in terms of genuinely many-atom interactions involving simultaneously all the near neighbours of an atom, up to 12 in number for a close packed structure? In fact the series (1.1) diverges, as we show dramatically in one simple case (§3).

Section 2 outlines some simple theoretical ideas about many-atom interactions but three basic points are in order here. Firstly, we know from the Uncertainty Principle of quantum mechanics that the bonding electrons will minimize their kinetic energy by spreading themselves out in molecular orbitals extending over as many atoms as possible (though this is balanced against other terms in the total energy). Thus many-atom forces should not be unexpected and we see that they, have their origin in the essential quantum nature of electrons. Secondly in nearly-free-electron metals a large item in the total energy of the system is the Fermi kinetic energy of the electron gas which depends only on its volume per atom. Now to define the Wigner–Seitz polyhedron around an atom and hence its volume, one needs to specify all the near neighbours. This points more specifically towards many-atom interactions, with 'many' of order the coordination number C. Thirdly there is the point about diamond above. In fact it was the kinetic energy which Mott & Jones noted was equal to the free-electron value if one took the measured band width from the bottom of the occupied band to the middle of the band gap, as seen in soft X-ray spectra. In conclusion, it seems that a series like (1.1) will not be a good and rapidly convergent representation of the binding energy to cover a wide range of coordination number and bonding geometries.

Section 4 mentions some evidence for the complexity of bonding in condensed matter, including an analysis of some *ab initio* calculations designed specifically to shed light on the many-atom character. Section 5 will review briefly two quite different approaches to what one does about it in computer simulations, and draw the conclusions together.

2. Some theory relating to many-atom interactions

The origin of many-atom interactions can be seen both from nearly-free-electron model and from the tight binding or linear combination of atomic orbitals (LCAO) model, which represent two opposite extreme views of electronic structure.

The subject has been reviewed recently (Heine & Hafner 1991) so that we will not discuss the nearly-free-electron approach here at length. The basic point has already been made in §1, namely the kinetic energy is a volume force and it requires a simultaneous specification of all the near neighbours to define the volume occupied by an atom. The implications of this for the structures of bulk solids and generally the success of this approach have been well documented (Heine & Weaire 1970; Hafner 1987).

The argument applies just as much to a semiconductor such as diamond, as it does to an sp bonded metal, because the occupied band width in the semiconductor is just governed by the free-electron value. This is completely consistent with the fact that the valence electrons in diamond are not uniformly spread out in the manner of a free-electron gas but are located almost entirely in the bonds between atoms, with almost zero electron density in the large holes between the bonds. It is easiest to understand this from a simple one-dimensional example of a chain of atoms with spacing a. At the Brillouin zone boundary we have free electron waves

$$\exp(\mathrm{i}\pi x/a) \quad \text{and} \quad \exp(-\mathrm{i}\pi x/a) \tag{2.1}$$

with kinetic energy

$$E_{\text{kin}} = h^2\pi^2/(2ma^2) \tag{2.2}$$

Now as is well known, Bragg reflection results in a mixing of these two waves, giving as wave function $\sin(\pi x/a)$ [or $\cos(\pi x/a)$]. This sine wave function has charge heaped up around positions $x = (n+\frac{1}{2})\,a$ where n is integer, so that we may describe it as a p-bonding orbital between all the atoms. But the relevant point here is that the kinetic energy of the sine state is still exactly equal to the free-electron value (2.2) as before. There is nothing inconsistent with the Uncertainty Principle in this because the heaping up of charge is formed from quantum interference between waves of the same kinetic energy: it does not require any extra kinetic energy. (I am indebted to Dr J. C. Phillips for this argument.)

Unfortunately the picture of volume forces derived from the free-electron gas model does not help us much in very inhomogeneous situations of low coordination number such as at surfaces or defects. Here we must turn to the opposite model of atomic orbitals. The result is that the formation energy varies with coordination number C roughly as

$$U = -(\text{const.})\,C^{\frac{1}{2}}. \tag{2.3}$$

It may be useful to rehearse in the simplest possible way the derivation of this result (Heine 1980). Consider a structure in which all atoms are geometrically equivalent with all nearest neighbours at the same distance, and through the device of periodic boundary conditions let us deal for mathematical convenience with finite system of N atoms instead of an infinite one. We assume a single s-orbital on each atom, though this can easily be generalized to a set of three p-orbitals or five d-orbitals (for each spin direction). Let the hopping integral or bonding energy between neighbouring atoms be h (negative), and take the self-energy of the orbital E_0 on a free atom as the zero of energy. The hamiltonian matrix H_{rs} describing the motion of the electrons then consists of zeros, except where r and s are neighbouring atoms when the matrix element is equal to h. We now consider the matrix H^2 and take its trace, i.e. the sum of the diagonal elements

$$\operatorname{Tr} H^2 = \sum_r (H^2)_{rr} = \sum_r \sum_s H_{rs} H_{sr} = -NCh^2. \tag{2.4}$$

We also apply a unitary transformation to H to turn it into diagonal form H' with diagonal elements equal to the eigenvalues E_n, which gives

$$\operatorname{Tr}(H')^2 = \sum_n E_n^2 = N(W_{\text{rms}})^2. \tag{2.5}$$

The eigenvalues E_n will be spread out into a band with density of states $n(E)$ per unit energy range (figure 1), and W_{rms} in (2.5) is defined to be the root-mean-square width of this distribution. By a simple theorem the trace is unchanged by a unitary transformation so that (2.4) and (2.5) are equal, which gives

$$W_{\text{rms}} = |h|\,C^{\frac{1}{2}}. \tag{2.6}$$

As already mentioned, the result can be generalized to any band of equivalent orbitals in the simplest two-centre approximation. For example for a d-band the $|h|$ in (2.6) has to be replaced by

$$[(\mathrm{dd}\sigma)^2 + 2(\mathrm{dd}\pi)^2 + 2(\mathrm{dd}\delta)^2]^{\frac{1}{2}}, \tag{2.7}$$

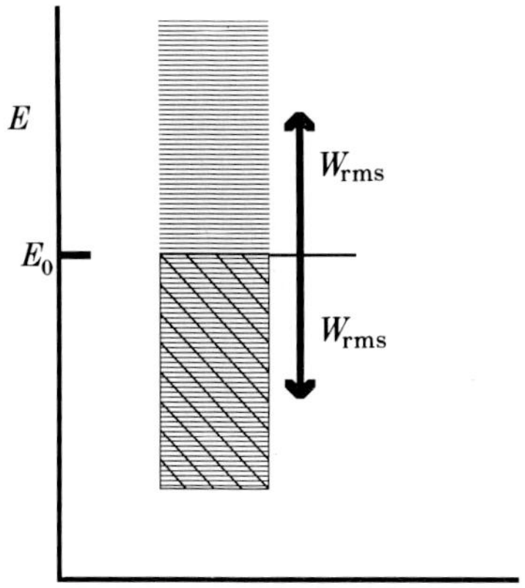

Figure 1. Electron energy levels of a system with a large number of atoms, spread into a band of root-mean-square width W_{rms}. For a half-filled band the shaded states are occupied by electrons.

where the (ddv) are the usual bonding matrix elements. The results (2.6), (2.7) are exact for any angular distribution of the neighbours about an atom.

To complete the argument we take one further, more approximate step. Let us suppose the band is half full. Then the occupied states have on average an energy about W_{rms} below the centre of the band, i.e. below the atomic level E_0. Thus the formation energy is

$$U(\text{per electron}) \approx -W_{rms}, \tag{2.8}$$

which together with (2.6) then yields (2.3). The precise value in (2.8) depends of course on the detailed band shape. The argument can be extended to some other degree of band filling, say one-third or three-quarters (Woolley 1990), but becomes invalid for nearly full or nearly empty bands where the band shape dominates. Our simple treatment also breaks down for a mixed s and p band, etc., though even there the variation with C seems to apply remarkably well in the case of one set of *ab initio* calculations (§4).

3. The non-convergence of *n*-atom interactions

In §2 we saw that the binding energy is roughly proportional to $C^{\frac{1}{2}}$ and in this section we investigate whether such a function can be represented as a convergent sum of two-, three-, n-atom interactions in the manner of equation (1.1). We are considering only the (negative) contribution to the formation energy of the solid from the bonding by the electrons due to the spread of energy levels into a band as in §2. Moreover we will keep the interatomic distance and hence the h in (2.6) constant. We continue with the model of a band of s-orbitals so that the two-atom interaction should be proportional to the coordination number C. Similarly we assume that an n-atom I_n interaction will be proportional to C^{n-1}. Under these conditions we write:

$$I_n = A_n C^{n-1}. \tag{3.1}$$

This is not quite correct. If there are C neighbours, the number of contributions of pairs of neighbours, i.e. three-atom combinations, is

$$\tfrac{1}{2}C(C-1). \tag{3.2}$$

Table 1. *Coefficients a_n in the least squares fit in equation (3.5), and root-mean-square error e_{rms} of the fit*

(Note that the coefficients have been severely rounded for presentation.)

n_{max}	2	3	4	5	6	7	8	9
a_2	1.2	2.1	2.9	3.6	4.4	5	6	7
a_3	—	−1.1	−3.8	−8.4	−15.1	−25	−39	−56
a_4	—	—	2.0	10.1	31.3	76	160	304
a_5	—	—	—	−4.3	−29.7	−119	−365	−936
a_6	—	—	—	—	10.5	92	459	1690
a_7	—	—	—	—	—	−28	−299	−1770
a_8	—	—	—	—	—	—	79	994
a_9	—	—	—	—	—	—	—	−231
e_{rms}	0.14	0.06	0.03	0.02	0.01	0.01	8.10^{-3}	6.10^{-3}

But this has C^2 as leading term so that our assumption (3.1) is approximate but sufficient for present purposes; and similarly for higher order interactions up to order $\frac{1}{2}C$.

The main purpose therefore is to test whether the form (2.3) can be expanded as follows

$$-|h|\, C^{\frac{1}{2}} = A_2\, C + A_3\, C^2 + \ldots + A_n\, C^{n-1} + \ldots. \tag{3.3}$$

It is convenient now to rewrite this in a dimensionless version in terms of the scale variable

$$z = \tfrac{1}{12}C \tag{3.4}$$

so that z runs from 0 to unity. Then (3.3) becomes

$$z^{\frac{1}{2}} = \sum_{n=2}^{n_{max}} a_n\, z^{n-1}, \tag{3.5}$$

where n_{max} is the cut-off on the series. We have done a least-squares fit to (3.5) for n_{max} ranging from 2 to 11, with the results given in table 1. The failure of (3.5) to give a convergent series is spectacular! Since $z^{n-1} = 1$ at the end point $z = 1$ for all n, the coefficients a_n give a direct measure of the magnitude of the corresponding interaction. For $n_{max} = 9$, we have $a_7 = -1770$! Moreover it is clear that as n_{max} is increased, the lower-order coefficients do not converge to some limiting value but continue to drift. it is noticeable that successive terms always alternate in sign throughout table 1 with each additional term trying to correct for the overshoot of the preceding one, a familiar behaviour of non-convergent series. We conclude that the function $C^{\frac{1}{2}}$ cannot be represented satisfactorily, even over a limited range, by a power series like (3.3).

Can we conclude from this that a series in n-atom interactions I_n (1.1) is similarly not convergent? At this point the argument becomes less than rigorous, but it suggests that we can so conclude at least in a practical sense. It certainly seems that we cannot hold to our cherished picture of two-atom terms dominating with some three-atom bond–bond additions and perhaps very small four-atom corrections. Look at $n_{max} = 3$ and 4 (table 1): the three-atom term has increased three-fold in magnitude as we add a four-atom correction, which even alters the two-atom term by 39%. Also the three-atom term has become larger than the two-atom one, which cannot be right. Moreover, the value of the two-atom term is double what it was for

$n_{\max} = 2$. And when we go one step further to $n_{\max} = 5$ the coefficients already start to go haywire. Of course the function $C^{\frac{1}{2}}$ is only a crude approximation but it captures the essence of metallic bonding quite well for the theoretical reasons of §2 and the empirical experience of §§4 and 5.

There is an alternative way of analysing the bonding energy (3.3) following Pauling's ideas. The energy of an electron in a bond between two atoms is $|h|$, and since there are more bonds than electrons we can think of the latter as having an energy $-|h|$ each. In fact there are very many ways of placing the set of electrons in the available bonds, with a 'resonance' between all of these configurations and a resulting 'resonance' energy lowering E_{res}

$$-|h|^{\frac{1}{2}} = -|h| + E_{\text{res}}. \tag{3.6}$$

We can now try to expand E_{res} in two-atom, three-atom, etc., terms analogous to (3.3) which (remembering (3.4)) leads to the series

$$z^{\frac{1}{2}} = \tfrac{1}{12} + \sum_{n=2}^{n_{\max}} b_n z^{n-1} \tag{3.7}$$

instead of (3.5). We have carried out at least square fit of (3.7), which turns out just as unsatisfactory as (3.5) and table 1.

Of course a function can be expanded as a Taylor series about some chosen point, and the result made to appear (spuriously, we believe) as a series of n-atom interactions. Suppose we consider the general function $F(C)$ and that in some situation only the values C_0 and $C_0 \pm 1$ occur. A good approximation should be the linear term

$$\begin{aligned} F(C) &\approx F(C_0) + (C - C_0) F'(C_0) \\ &= [F(C_0) - C_0 F'(C_0)] + C F'(C_0) \\ &= A_1 + A_2 C. \end{aligned} \tag{3.8}$$

Hence if one sweeps the constant A_1 under the carpet, one can appear to have a two-atom interaction model. More generally, one may be able to obtain simple models to fit the bonding over a limited range of geometries, but it is probably spurious to extrapolate with them to the bonding in very different situations. And that is the task we face in the computer simulation of complex atomic processes, as emphasized in §1: these almost always involve some very non-standard arrangement of atoms.

4. Evidence for many-atom forces

The existence of volume forces in metals has been discussed since the beginning of quantum mechanics as the origin of the Cauchy anomaly in the elastic constants. The importance of such volume forces for the equilibrium volumes and crystal structures of sp-bonded elements is also well known (Heine & Weaire 1970, particularly pp. 259–282; Hafner & Heine 1983). Recently a whole conference was devoted to a discussion of many-atom forces (Nieminen *et al.* 1990). The success of the 'glue' models (see next section), which take many-atom forces specifically into account, has been impressive particularly in accounting for subtle effects such as the reconstructions of the Au surfaces.

By way of contrast, one can point to the sad history of attempts to find a satisfactory force model for Si. There has been a whole succession of models, each

Table 2. *Calculated formation energy U of various structures spanning a range of coordinated numbers*

structure	coordination	energy (atom eV)	U eV
atom	0	−54.95	0
line	2	−56.28	−1.33
graphite mesh	3	−56.95	−2.00
diamond	4	−57.42	−2.47
square mesh	4	−57.29	−2.34
square bilayer	5	−57.64	−2.69
simple cubic	6	−57.91	−2.96
triangular mesh	6	−57.49	−2.54
vacancy lattice	8	−58.10	−3.15
face centred cubic	12	−58.31	−3.36

fitted to experimental data plus the computed energies of a range of unstable structures covering quite a variety of bonding geometries. Yet each model failed when confronted a year or so later by the energy of some new structure. The latest chapter in this saga comes from the quantum calculation of the energies, structures and dissociation channels of the cluster molecules Si_N up to $N = 11$. The comparison between the predictions of the best empirical models with the fully quantum calculations can only be described as abysmal (Andreoni & Pastore 1990). These models are all based on n-atom expansions of type (1.1).

We would like to present here the results of some recent calculation on aluminium. These are *ab initio* quantum mechanical calculations of 10 structures with the number of nearest neighbours varying from $C = 0$ to the normal FCC structure of the bulk metal with $C = 12$. In all cases the energy was calculated keeping the nearest neighbour distance fixed, equal to the equilibrium value for the bulk metal. The results therefore focus attention directly on the variation of the energy with the coordination C. For $C = 2$ we have a linear chain of atoms. A single layer with the graphite structure has $C = 3$, while a layer with a square mesh has $C = 4$. The three-dimensional diamond structure also has $C = 4$. The $C = 5$ is obtained as a double layer, i.e. two square layers with the atoms of one over the other. $C = 6$ is achieved in two ways, one being a triangular layer and the other the simple cubic structure. A closely spaced lattice of vacancies in the FCC structure has $C = 8$. In each case all atoms are equivalent by symmetry. The calculations were carried out using a pseudopotential for aluminium (Goodwin *et al.* 1990) and the method described in §5. The chain and layer structures were modelled in a supercell with an appreciable amount of free space around them, repeatedly periodically in three dimensions. The wave functions were expanded in plane waves with a cut-off of 190 eV. A large number of k-points were sampled in the Brillouin zone using the KP scheme (Robertson & Payne 1990). The energy of an isolated atom was calculated in a supercell using the same program in an artificial spherically symmetric configuration with all six 2p spin-orbitals occupied equally, which corresponds to a 'metallic' state with $C = 0$ for comparison with the other calculations (table 2). A more detailed discussion of these results will be given elsewhere (Robertson *et al.* 1991).

The results are given in table 2 and figure 2. They are offered as a test-bed for the development of empirical potentials and study of the variation with C.

We will only give the very simplest analysis here. As a very crude approximation,

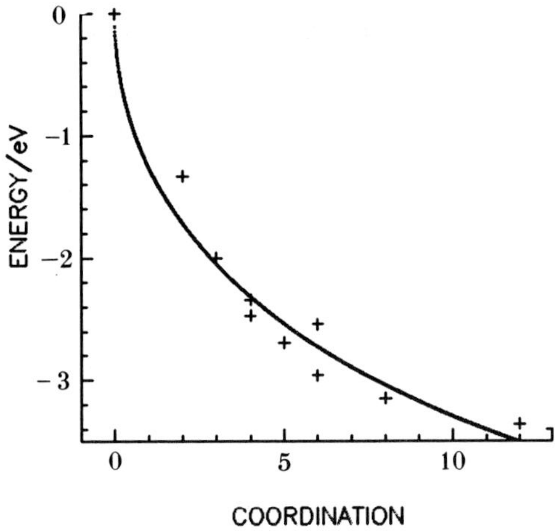

Figure 2. The formation energy U from table 2 (crosses) as a function of the coordination number C. The line is a least-squares fit of the form of equation (4.1).

we expect the formation energy $U(C)$ to consist of two parts, (i) a short range repulsion and (ii) the bonding by the electrons. The first will be proportional to C for fixed nearest neighbour distance: it is due to the coulomb repulsion of the nuclei and the exclusion principle in the overlap of the atoms. For the bonding we use (2.3). We have therefore done a least-squares fit to the energies using

$$U(C) = aC - bC^{\frac{1}{2}}, \tag{4.1}$$

where $U(C)$ is the total energy of the system per atom relative to that of the free atom ($C = 0$) defined as above.

The result of the fit is shown in figure 2 and is seen to describe the overall variation with C surprisingly well. We interpret this as general qualitative support for the picture of many atom forces painted in §2. The discussion there is only intended as a broad brush treatment, to suggest that the correct starting point is a formulation in terms of many-atom forces as in the 'glue' models mentioned in §5 rather than the series (1.1). Of course we are very conscious of the shortcomings of (4.1) as a theoretical model. It neglects the effect of second neighbours which may account for the energy differences between two pairs with the same C. We also note that (2.3) does not apply to a mixed band of s and p electrons such as we have in aluminium, and in any case ignores the shape of the band.

5. Implications for simulations

This section indicates two approaches that are being used to address the problem of many-atom interactions in computer simulations, and then add a few thoughts for the future.

The first approach consists of setting up models of the many-atom interactions, which we refer to collectively as 'glue' models. The formation energy is written

$$U = \sum_{i} U_e(C_i) \sum_{i \neq j}\sum \Phi(R_{ij}). \tag{5.1}$$

Here i and j denote the atoms, C_i the coordination number of atom i, and Φ is a pairwise interaction depending on the interatomic distance R_{ij}. the $U_e(C)$ is a bonding or 'glue' or 'embedding' energy depending on the coordination number C in the spirit of §§2 and 4. However we need a precise definition

$$C_i = \sum_j c(R_{ij}) \tag{5.2}$$

of the coordination number. The function $c(R)$ is normally chosen so that it is simply unity for a neighbour j at the equilibrium distance of the nearest neighbours R_{nn} in the bulk material. The function $C(R)$ is monotonically decreasing so that a neighbour closer than R_{nn} counts a little more than unity and neighbours more distant than R_{nn} count less, with $c(R)$ being cut off to zero at some suitable distance around the second neighbours. There are various versions of this basic approach. The Trieste group seems to have invented the expression 'glue model' to signify that the electron gas is a ubiquitous glue holding the structure together with a strength U_e depending on its density (Ercolessi *et al.* 1988). There and in the 'embedded atom' method (Foiles *et al.* 1986) all three functions $U_e(C)$, $c(R)$ and $\Phi(R)$ are fitted purely empirically to a variety of experimental data, whereas Finnis & Sinclair (1984) used the square-root form (2.1) for $U_e(C)$. The least empirical is the 'effective medium theory' (Jacobsen *et al.* 1987) where the embedding function $U_e(C)$ is computed quantum mechanically by inserting an atom in a free electron gas of density ρ. The ρ is determined also by the quantum calculation from the electron density contributed by neighbouring atoms. In other words in the interaction of an atom with its neighbours, the latter are replaced by a uniform electron gas whose density is determined selfconsistently.

What is common to all these glue formulations is that the embedding function U_e is usually negative with an upward curvature like (2.3) that gives the model its characteristically 'metallic' behaviour. Metals are usually stated to prefer closely packed structures, but a more apt description is to say that the ion cores swimming in a sea of electrons want to space themselves out as uniformly as possible for a given overall atomic density. This principle in three dimensions leads to the FCC structure, for example, and in two dimensions to the triangular close packed mesh. Now consider the (110) surface of gold. In its ideal (unreconstructed) state it has atoms with coordination $C_1 = 7$ in the surface layer and $C_2 = 11$ in the second layer. For the (111) surface we have $C_1 = 9$ and C_2 is already the bulk value 12. Because of the upward curvature of $U_e(C)$, we have

$$U_e(9) < \tfrac{1}{2}[U_e(7) + U_e(11)]. \tag{5.3}$$

This means that it is energetically favourable to replace pairs of atoms with $C = 7$ and 11 by atoms with $C = 9$, i.e. to replace (110) surface by 'metallic' close packed (111) surface. And indeed the observed 2×1 and 3×1 'missing row' reconstruction of the Au (110) surface has been described in terms of forming small facets of (111) surface. Of course, the detailed energetics are more complicated than equation (5.3), involving the densities of surface atoms per unit area and the bonding of the atoms at the ridges and furrows of the reconstruction. But we believe our argument captures the essence of what drives the reconstruction. The story is similar on the (100) surface where $C_1 = 8$ and $C_2 = 12$. The number of atoms per unit area is less than for the close packed (111) surface. Thus to turn (100) surface into (111) surface means taking atoms of $C = 8$ with some of $C = 12$ and replacing then by atoms with $C = 9$. An equation similar to (5.3) again suggests this may be energetically favourable and indeed the (100) surface reconstructs to give a close packed (111)-type monolayer lying on top of the (100) structure. Well, these are very hand-waving arguments but it is appealing to have some broad understandable qualitative mechanism. We refer to pp. 82–83 of Heine (1980) for some further examples based on the same curvature argument.

However, it must be admitted that none of these glue models seem to have been tested very rigorously, for example over a wide range of C or R. *Ab initio* calculations of the type discussed in §4 may have a contribution to make here, in augmenting experimental data by energy measurements in 'computer experiments'. A further shortcoming is the absence of any angular forces which certainly will not suffice for covalent semiconductors or transition metals. Of course some people would regard it as inappropriate to treat silicon starting with a glue model based on volume forces, but our arguments of §2 would suggest that to be quite sensible with the addition of angular forces. See also Heine & Hafner (1991) for a discussion of the relative magnitudes of the volume, pairwise an angular terms in silicon.

We now turn to the second approach to complicated interatomic forces in a simulation. This is to hit it with an *ab initio* quantum calculation for the whole system. With current supercomputers one can do this for a simulation of modest size (Ihm 1988; Payne *et al.* 1991). The complex process to be simulated has to be dissected into its physically significant aspects. One of these at a time has to be modelled in a supercell of typically 10 to 100 atoms, repeated periodically in three dimensions. (A surface can be simulated by including some empty space in the supercell.) Static simulations usually focus on the energy, e.g. a barrier height for diffusion, or/and on relaxing the atoms step by step to find their equilibrium positions, e.g. at a surface reconstruction. Dynamic simulations include a stepwise integration of the equations of motion for the ion cores to represent thermal agitation. In most such work the atoms are represented by pseudopotentials so that one is only solving the quantum mechanics for the valence electrons. The wave functions are expanded typically in thousands of plane waves, while the electron exchange and correlation is treated by the local density approximation within the density functional formulation of the total energy. For a review, see Ihm (1988).

Although the *ab initio* quantum calculation gives completely the interatomic forces for any arrangement of atoms, the size of system that can be simulated is somewhat limited, as already mentioned. In this connection considerable progress is being made in speeding up the numerical algorithms, based on the innovative ideas of Car & Parrinello (1985) and subsequent developments (Payne *et al.* 1991).

However, even with the next generation of supercomputers and probably several generations after that, the maximum number of atoms in the unit cell will be a restriction on the quantum simulations. More work needs to be done to find hybrid methods where *ab initio* quantum calculations give information about the interatomic forces for critical geometries furthest removed from standard bonding arrangements, and this information then gets fed into large simulations of more traditional type. One might for instance generate a data set by quantum calculation analogous to that of §4, but selected specifically to cover the relevant geometries. One is then using the fitted model to interpolate from, rather than extrapolating from the fitting configurations. At the simplest level this has already been done for example to obtain the phonon free energies of some SiC polytypes with fairly large unit cells. First some meagre experimental data on phonon frequencies in cubic SiC (the simplest polytype) were supplemented by phonon frequencies and eigenvectors at other points in the Brillouin zone from 'frozen phonon' quantum calculations. The phonon spectrum of cubic SiC was then fitted by an extended shell model. In this context having the phonon eigenvectors from quantum calculations was very important because these are difficult to obtain from experiment and the frequencies alone are known not to determine the model uniquely (Cheng *et al.* 1989). Finally the extended shell model could be applied to the whole range of polytypes because they all have locally the same tetrahedral bonding as the cubic form (Cheng *et al.* 1990). There are other examples along similar lines, for instance calculations on point defects, and modelling the interactions between organic molecules.

In conclusion, the development of reliable quantum calculations for large systems is overcoming the lack of adequate information in traditional simulations about interatomic bonding energies in unconventional geometries. On the other hand models are being developed to represent explicitly the many-atom character of interatomic forces, at least to some degree. The hope is that these can be developed further, and used to transfer the bonding information from relatively modest *ab initio* calculations to large simulations.

References

Andreoni, W. & Pastore, G. 1990 (In the press.)

Car, R. & Parrinello, M. 1985 *Phys. Rev. Lett.* **55**, 2471–2474.

Cheng, C., Kunc, K. & Heine, V. 1989 *Phys. Rev.* B **39**, 5892–5898.

Cheng, C., Heine, V. & Jones, I. L. 1990 *J. Phys. Condens. Matter* **2**, 5097–5114.

Ercolessi, F., Parrinello, M. & Tosatti, E. 1988 *Phil. Mag.* A **58**, 213–226.

Finnis, M. W. & Sinclair, J. E. 1984 *Phil. Mag.* A **50**, 45–55.

Foiles, S. M., Baskes, M. I. & Daw, M. S. 1986 *Phys. Rev.* B **33**, 7983–7991.

Goodwin, L., Needs, R. J. & Heine, V. 1990 *J. Phys. Condens. Matter* **2**, 351–366.

Hafner, J. 1987 *From hamiltonians to phase diagrams.* Berlin: Springer-Verlag.

Hafner, J. & Heine, V. 1983 *J. Phys.* F **23**, 2479–2501.

Heine, V. 1980 *Solid St. Phys.* **35**, 1–127.

Heine, V. & Jones, R. O. 1969 *J. Phys.* C **2**, 719–732.

Heine, V. & Hafner, J. 1991 In *Many-atom interactions in solids* (ed. R. Nieminen, M. J. & M. Manninen), pp. 12–33. Berlin: Springer-Verlag.

Heine, V. & Weaire, D. 1970 *Solid St. Phys.* **24**, 249–463.

Ihm, J. 1988 *Rep. Prog. Phys.* **51**, 105–142.

Jacobsen, K. W., Nørskov, J. K. & Puska, M. J. 1987 *Phys. Rev.* B **35**, 7423–7442.

Mott, N. F. & Jones, H. 1936 *The theory of the properties of metals and alloys*, p. 159. Oxford: Clarendon Press.

Nieminen, R., Puska, M. J. & Manninen, M. (eds) 1990 *Many-atom interactions in solids.* Berlin: Springer-Verlag.

Payne, M. C., Allen, D. C., Teter, M. P. & Joannopolous, J. 1991 *Rev. mod. Phys.* (In the press.)

Robertson, I. J. & Payne, M. C. 1990 *J. Phys. Condens. Matter* **2**, 9837–9852.

Robertson, I. J., Payne, M. C. & Heine, V. 1991 (Submitted.)

Woolley, R. G. 1990 *J. Phys.* B **23**, 1563–1576.

Discussion

J. N. Murrell (*University of Sussex, U.K.*). I have for many years been working on potential energy functions for polyatomic molecules; functions that can be used both for spectroscopy and chemical kinetics (bond breaking). Only recently have I applied some of these ideas to solids and so far with any quantitative study, only to the group IV solids. On the basis of my current results I believe the prognosis from many-body expansion of the potential to be much better than suggested by the speaker. In short, I have a two-body plus three-body potential for Si which reproduces the lattice energy, lattice spacing, elastic constants, and phonon-dispersion curves with good accuracy and gives sensible energies for other bulk phases and for a graphitic structure. It may be that this potential fails for other features, e.g. surface reconstruction, and performs badly for melting, etc. With my present knowledge, however, I believe the potential is better than any other that has been published.

V. Heine. Although I am pessimistic about the success of Professor Murrell's approach, such success would be very important and I look forward to his further results. Even partial success could be very useful, for example if it was sufficient for all likely configurations of atoms at defects in the bulk solid.

J. C. Phillips (*AT&T Bell Labs, Murray Hill, U.S.A.*). Has Professor Heine extended his results to Si as well as Al?

V. Heine. No, we have not calculated our structures for Si.

J. C. Phillips. We believe that the many-atom force model of Chelikowsky at present gives the best available account of structural trends of Si clusters up to 30 atoms, although it of course does not include the Jahn–Teller distortion energies which become small for clusters containing more than 10 atoms.

D. Weaire (*Trinity College, Dublin, The Republic of Ireland*). Professor Heine pictured an atom in the solid state as a piece of jellium of volume defined by its Wigner–Seitz cell and hence having a *volume*-dependent energy. This defines a model, familiar enough in the case of a perfect crystal, but has it ever been used to calculate, say, dynamical properties?

V. Heine. I do not know of any detailed work along such lines. It would be quite possible for the phonons in the perfect bulk metal and for some defects perhaps. However, the volume would not be definable this way at a surface where it would be infinite, and I have doubts about its value even at a vacancy.

Electronic fluctuation and the van der Waals metal

BY N. W. ASHCROFT
Laboratory of Atomic and Solid State Physics, Cornell, University, Ithaca, New York 14853–2501, U.S.A.

Interactions determining the structure of condensed matter can systematically be developed starting with the fundamental view of such systems as neutral canonical ensembles of nuclei and electrons and proceeding to the more common viewpoint of ions and valence electrons but retaining in both the dominant fluctuational effects normally omitted.

---

1. Introduction

The purpose of this paper is to show that on energy scales apposite to structure in metals, bonding energies associated with electromagnetic fluctuation can be significant. In lowest order these fluctuational effects are almost universally referred to van der Waals interactions (the neutral atom case with retardation effects neglected was first worked out by London (1930)). The terminology will be extended here to systems possessing charge which is free in the normal electromagnetic sense. The issue is therefore the degree to which fluctuations in both bound and free charge can be manifested in bonding, structure, and in ordering.

The importance of electromagnetic fluctuation in the cohesive properties of the metallic state has earlier been emphasized by Rehr *et al.* (1975) and in a detailed proposal for pair interactions by Mahanty & Taylor (1978) and by Mon *et al.* (1979). The subject has recently been reviewed by Barash & Ginzburg (1988) with a focus on overall energetics. In what follows fluctuational effects are examined as a possible source of correction on energy scales that in some cases can rival those that are associated with standard mean-field-based interactions.

2. Condensed matter as a two-component system

The emphasis is to be placed on measurable structure, and it is therefore necessary to introduce from the outset a notation which links immediately to observable structural quantities. These are the one- and two-particle densities for electrons and nuclei respectively; later it will be necessary to introduce the n-particle generalizations for the composite objects referred to normally as atoms or ions. For convenience the discussion begins with a macroscopic volume Ω of an elemental system, atomic number Z_A, with N_n (10^{23}) nuclei of mass m_n having instantaneous coordinates $\boldsymbol{r}_{n1}, \ldots, \boldsymbol{r}_{nN_n}$. In a neutral canonical ensemble, there will also be present $N_e = Z_A N_n$ electrons, of mass m_e having instantaneous coordinates $\boldsymbol{r}_{e1}, \ldots, \boldsymbol{r}_{eN_e}$. For particles of type α ($\alpha = e, n$) the one particle density operator is

$$\hat{\rho}_\alpha^{(1)}(\boldsymbol{r}) = \sum_i^{N_\alpha} \delta(\boldsymbol{r} - \boldsymbol{r}_{\alpha i}) \tag{1}$$

Phil. Trans. R. Soc. Lond. A (1991) **334**, 407–423
Printed in Great Britain

whose quantum statistical average over states of the entire system is

$$\rho_\alpha^{(1)}(\boldsymbol{r}) = \langle \hat{\rho}_\alpha^{(1)}(\boldsymbol{r}) \rangle \tag{2}$$

and it has the meaning that $\rho_\alpha^{(1)}(\boldsymbol{r})\,\mathrm{d}\boldsymbol{r}$ is the probability of finding, at a given instant, a particle of type α in $\mathrm{d}\boldsymbol{r}$ at $\boldsymbol{r}$. At the level of electrons and nuclei (Chihara 1985) there are at most two-particle interactions; accordingly the two-particle density operator is also introduced

$$\hat{\rho}_{\alpha\beta}^{(2)}(\boldsymbol{r},\boldsymbol{r}') = \hat{\rho}_\alpha^{(1)}(\boldsymbol{r})\,\hat{\rho}_\beta^{(1)}(\boldsymbol{r}') - \delta_{\alpha\beta}\,\hat{\rho}_\alpha^{(1)}(\boldsymbol{r})\,\delta(\boldsymbol{r}-\boldsymbol{r}'), \tag{3}$$

whose corresponding average is

$$\rho_{\alpha\beta}^{(2)}(\boldsymbol{r},\boldsymbol{r}') = \langle \hat{\rho}_{\alpha\beta}^{(2)}(\boldsymbol{r},\boldsymbol{r}') \rangle, \tag{4}$$

which has the meaning that $\rho_{\alpha\beta}^{(2)}(\boldsymbol{r},\boldsymbol{r}')\,\mathrm{d}\boldsymbol{r}\,\mathrm{d}\boldsymbol{r}'$ is the probability of finding particles of type α and β simultaneously in $\mathrm{d}\boldsymbol{r}$ and $\mathrm{d}\boldsymbol{r}'$, which are themselves separated by $\boldsymbol{r}-\boldsymbol{r}'$. For the length scales appropriate to the condensed state of matter, the nuclei are taken as point objects.

The problem is simplified in an inessential way by viewing the system as non-relativistic. If $v_\mathrm{c}(r) = e^2/r$ is the basic Coulomb interaction then the hamiltonian of system of electrons and nuclei can be written down, using (4) as

$$H = \sum_\alpha \left\{ \hat{T}_\alpha + \tfrac{1}{2}\sum_\beta \int_\Omega \mathrm{d}\boldsymbol{r} \int_\Omega \mathrm{d}\boldsymbol{r}'\, v_\mathrm{c}(\boldsymbol{r}-\boldsymbol{r}')\, Z_\alpha Z_\beta\, \hat{\rho}_{\alpha\beta}^{(2)}(\boldsymbol{r}-\boldsymbol{r}') \right\}, \tag{5}$$

where $Z_\mathrm{e} = -1$ and $Z_\mathrm{n} = Z_\mathrm{A}$. Here $\hat{T}_\alpha = (-\hbar^2/2m_\alpha)\Sigma_i \nabla_{\alpha i}^2$ are the kinetic energies; there is no explicit reference to spin. The notation introduced readily permits extension from elements to alloys or compounds.

3. Reduced hamiltonian

For elements with a relatively small number of electrons per nucleus it is now becoming possible to apply many-body quantum Monte-Carlo simulation techniques directly to (5) (Sugiyama *et al.* 1989). But with the possible exception of hydrogen ($Z_\mathrm{A} = 1$) (Ashcroft 1981), or with the exception of elements placed under conditions so extreme that all $N_\mathrm{n} Z_\mathrm{A}$ electrons are unbound, a description of condensed matter proceeding from (5) is not entirely appropriate. Rather (5) is modified to reflect what is known already from atomic physics, namely, that under ordinary conditions there is considerable condensation or agglomeration of electronic charge around the nuclei. In particular, for vapour phases of (5) the electronic density $\rho_\mathrm{e}^{(1)}(\boldsymbol{r}) = \langle \hat{\rho}_\mathrm{e}^{(1)}(\boldsymbol{r}) \rangle$ is extremely close to that of essentially isolated atoms. Such structure as $\rho_\mathrm{e}^{(1)}(\boldsymbol{r})$ possesses is just the 'shell-structure' of atomic physics (though $\rho_\mathrm{e}^{(1)}(\boldsymbol{r})$ is in fact remarkably devoid of structure) and, extreme conditions excepted, much of this shell structure is immune to passage of the element from vapour to condensed state. It is therefore useful to make a division of $\hat{\rho}_\mathrm{e}^{(1)}(\boldsymbol{r})$ into contributions arising from electrons that will exist in states dominating most aspects of the physics of condensed matter, and the remainder to be found in states bound at energies far larger than are characteristic of the condensed state. The exchange terms introduced by site identification can be incorporated within approximations that amount to a redefinition of the primary interaction, and especially its high q components (since exchange manifests itself mainly at short range). This point of view will be adopted below, in (11*a*).

The notion of valence, as applied to condensed matter must clearly be state dependent since site localization carries with it a presumption that orbitals exist whose range or scale is known. If external conditions are later imposed that require nuclei to approach closer than the diameter of such orbitals, then the system will be brought close to and beyond the onset of a Mott transition for this set of electrons. It is then appropriate to include them in the valence set, the next lower orbitals (and those below them) now constituting the localized charge. The general concept of electron-derived interactions is therefore necessarily a function of density even though for many applications this state dependence can plausibly be ignored.

With these restrictions in mind the one-particle electronic density operator is written

$$\hat{\rho}_{\mathrm{e}}^{(1)}(\boldsymbol{r}) = \sum_{i=1}^{ZN_{\mathrm{n}}} \delta(\boldsymbol{r}-\boldsymbol{r}_{\mathrm{e}i}) + \sum_{j} \sum_{i=1}^{Z_{\mathrm{c}}} \delta(\boldsymbol{r}-\boldsymbol{r}_{\mathrm{n}j}-\boldsymbol{r}_{\mathrm{e}i}^{j}), \tag{6}$$

where a number Z_{c} (of the Z_{A} per nucleus) electrons is to be in the localized (or 'core') class as described above, and the remainder $Z = Z_{\mathrm{A}} - Z_{\mathrm{c}}$ per nucleus are taken as valence electrons. If static conduction is eventually found to take place in the states of H, the processes will involve only the valence electrons. In what follows Z_{c}, Z (and obviously Z_{A}) are all taken to be conserved quantities.

By an ion is now meant the usual composite object formed by a nucleus together with Z_{c} electrons considered to reside in the localized states of H. To this ion can be assigned a site hamiltonian $\hat{h}_j$, which includes all electronic kinetic energies, all mutual Coulomb repulsions, and all Coulomb attractions with the nucleus. To explore the fluctuational aspects in detail, it is particularly expedient to introduce a *site density* operator

$$\hat{\rho}_j^{(1)}(\boldsymbol{r}) = Z\delta(\boldsymbol{r}-\boldsymbol{r}_{\mathrm{n}j}) + Z_{\mathrm{e}} \sum^{Z_{\mathrm{c}}} \delta(\boldsymbol{r}-\boldsymbol{r}_{\mathrm{n}j}-\boldsymbol{r}_{\mathrm{e}i}^{j}). \tag{7}$$

The meaning of $\hat{\rho}_j^{(1)}(\boldsymbol{r})$ is apparent from its Fourier representation

$$\hat{\rho}_j^{(1)}(\boldsymbol{q}) = \exp(\mathrm{i}\boldsymbol{q}\cdot\boldsymbol{r}_{\mathrm{n}j}) \left\{ Z + \sum_{i=1}^{Z_{\mathrm{c}}} (1-\exp(\mathrm{i}\boldsymbol{q}\cdot\boldsymbol{r}_{\mathrm{e}i}^{j})) \right\}, \tag{8}$$

which displays the number-fluctuations about the average valence Z. At long range $(q \to 0)$

$$e\hat{\rho}_j^{(1)}(\boldsymbol{q}) \approx \exp(\mathrm{i}\boldsymbol{q}\cdot\boldsymbol{r}_{\mathrm{n}j})\{Ze + \mathrm{i}\hat{d}_j\cdot\boldsymbol{q} + \ldots\}, \tag{9}$$

where relative to the nucleus at j,

$$\hat{d}_j = \sum_{i=1}^{Z_{\mathrm{c}}} (-e)\,\boldsymbol{r}_{\mathrm{e}i}^{j}$$

is the instantaneous dipole operator giving the leading multipole correction to the standard monopole, Ze, normally assumed for such an ion

The valence electron operator is defined by

$$\hat{\rho}_{\mathrm{v}}^{(1)}(\boldsymbol{r}) = \sum_{i=1}^{N_{\mathrm{v}}} \delta(\boldsymbol{r}-\boldsymbol{r}_{\mathrm{e}i}), \tag{10}$$

where $N_v = ZN_n$. With these definitions the original hamiltonian (5) can be rewritten in a form appropriate to a description of the metallic state, namely

$$H - \sum_{j=1}^{N_n} \hat{h}_j = \hat{T}_n + \frac{1}{2\Omega} \sum_{z,j,j'} \hat{\rho}_j^{(1)}(\boldsymbol{q})\,\hat{\rho}_{j'}^{(1)}(-\boldsymbol{q})\,\bar{v}(q) \tag{11a}$$

$$+ \hat{T}_e + \frac{1}{2\Omega} \sum_{q} (\hat{\rho}_v^{(1)}(\boldsymbol{q})\,\hat{\rho}_v^{(1)}(-\boldsymbol{q}) - N_v)\, v_c(q) \tag{11b}$$

$$- \frac{1}{\Omega} \sum_{q,j} \rho_v^{(1)}(\boldsymbol{q})\,\hat{\rho}_j^{(1)}(-\boldsymbol{q})\,\omega(\boldsymbol{q}), \tag{11c}$$

where $v_c(q) = 4\pi e^2/q^2$ and

$$\hat{\rho}_v^{(1)}(\boldsymbol{q}) = \sum_{i=1}^{N_v} \exp(\mathrm{i}\boldsymbol{q}\cdot\boldsymbol{r}_{ei}).$$

Here $\bar{v}(q)$ incorporates the corrections, to $v_c(q)$, for core–core exchange alluded to above. By an extension of those arguments, there is also core–valence exchange: further, in the eventual states of H, core and valence (localized and itinerant) states are orthogonal. Again, these effects are included at the level of the hamiltonian in (11 c) by the replacement of $v_c(q)$ by $w(q)$, a pseudopotential (Heine 1970). In general w is a non-local operator; for small q, $Z\omega(q) \approx Zv_c(q)$. The pseudopotential concept is by no means limited in its usefulness to simple sp metals, but has also been extended to transition metals by Moriarty (1990), and also applied in such systems to the determination of multi-centre potentials.

It is the manifestation of electromagnetic fluctuation associated with the reduced hamiltonian (11) that is of primary interest in what follows. For the most part, the length scales of interest are a few lattice constants and therefore retardation effects can be neglected. Observe that the reduced hamiltonian (11) can easily be extended to an alloy or compound metal by noting that the equivalent of (8) for component α will be

$$\hat{\rho}_{\alpha j}^{(1)}(\boldsymbol{q}) = \exp(\mathrm{i}\boldsymbol{q}\cdot\boldsymbol{r}_{n\alpha j})\left\{Z_\alpha + \sum_{i=1}^{z_{c\alpha}} (1 - \exp(\mathrm{i}\boldsymbol{q}\cdot\boldsymbol{r}_{ei}^{\alpha j}))\right\},$$

where Z_α is the nominal valence of each component. Charge transfer in compound or alloy formation is encompassed by assigning appropriate values to Z_α (negative values are permitted, for example). For either the single or multi-element case, (11 c) contains within it the physics of fluctuational attraction of an electron to a localized charge distribution and hence to the notion of electron affinity (or electronegativity).

If dipolar terms are taken to dominate the fluctuations (and hence in the expansion of (8)) then (11) becomes

$$H - \sum_{j=1}^{N_n} \hat{h}_j = \hat{T}_n + \frac{Z^2}{2} \int \mathrm{d}\boldsymbol{r} \int \mathrm{d}\boldsymbol{r}'\,\hat{\rho}_{nn}^{(2)}(\boldsymbol{r},\boldsymbol{r}')\,\bar{v}(\boldsymbol{r}-\boldsymbol{r}') + \tfrac{1}{2}\sum_{jj'}{}' \int \mathrm{d}\boldsymbol{r} \int \mathrm{d}\boldsymbol{r}'(\hat{d}_j\cdot\nabla_r)\,(\hat{d}_{j'}\cdot\nabla_{r'})\,\bar{v}(\boldsymbol{r}-\boldsymbol{r}') \tag{11d}$$

$$+ \hat{T}_v + \tfrac{1}{2} \int \mathrm{d}\boldsymbol{r} \int \mathrm{d}\boldsymbol{r}'\,\hat{\rho}_{vv}^{(2)}(\boldsymbol{r},\boldsymbol{r}')\,v_c(\boldsymbol{r}-\boldsymbol{r}') + \int \mathrm{d}\boldsymbol{r} \int \mathrm{d}\boldsymbol{r}'\,\hat{\rho}_n^{(1)}(\boldsymbol{r})\,\hat{\rho}_v^{(1)}(\boldsymbol{r}')\,\omega(\boldsymbol{r}-\boldsymbol{r}') \tag{11e}$$

$$+ \sum_j \int \mathrm{d}\boldsymbol{r}\,\hat{\rho}_v^{(1)}(\boldsymbol{r})\,(\hat{d}_j\cdot\nabla_f)\,\omega(\boldsymbol{r}-\boldsymbol{r}'). \tag{11f}$$

It should be apparent that correlation effects from on-site physics are essential to (11). In this respect both the description to follow, and its consequences, differ notably from the standard Hubbard model where site-fluctuations are eliminated from the outset. The manifestation of polarizable ions in effective interactions have previously been explored for ionic crystals in what is known as the shell model (Roberts 1950).

4. Mean-field effective interactions

Implicit in the introduction of the site density operator $\hat{\rho}_j^{(1)}(\boldsymbol{r})$ (see (7)) is the adibatic (Born–Oppenheimer) separation of timescales. For most purposes it is admissible to proceed on the solution of the states of (11) by solving an electron problem in which the nuclear densities $\hat{\rho}_{\mathrm{n}}^{(1)}(\boldsymbol{r})$, $\hat{\rho}_{\mathrm{nn}}^{(2)}(\boldsymbol{r},\boldsymbol{r}')$ are regarded as parameters. For the majority of electron states the adiabatic principle can be invoked (Moody *et al.* 1989). Within this assumption the notion of an effective hamiltonian controlling ionic motion can be developed by tracing out the electronic states (e) for each ionic or nuclear configuration (e(n)). The partition function for the system than has the form ($\beta = 1/k_{\mathrm{B}}\mathrm{T}$)

$$Z = Z(\Omega, T) = \mathrm{Tr}\,\mathrm{e}^{-\beta H} = \mathrm{Tr}_{\mathrm{n}} \exp\left(-\beta H_{\mathrm{eff}}(\boldsymbol{r}_{\mathrm{n}i}, \ldots, \boldsymbol{r}_{\mathrm{n}N_{\mathrm{n}}}; \Omega, \beta)\right) \tag{12}$$

with

$$H_{\mathrm{eff}} = -k_B T \ln \mathrm{Tr}_{\mathrm{e(n)}} \exp\left(-\beta H(\{\boldsymbol{r}_{\mathrm{e}i}; \boldsymbol{r}_{\mathrm{n}j}\})\right).$$

In this form, the effective hamiltonian would include, in principle, contributions of entropic origin from the electronic subsystem. For many applications, however, the chosen thermodynamic conditions are such that the electron system is very close to its ground state, $\Psi^0_{\mathrm{e(n)}}$ say, and if this is so the implied ground state trace leads to

$$H_{\mathrm{eff}}(\{\boldsymbol{r}_{\mathrm{n}j}\}; \Omega) = \hat{T}_{\mathrm{n}} + V(\boldsymbol{r}_{\mathrm{n}1}, \ldots, \boldsymbol{r}_{\mathrm{n}N_{\mathrm{n}}}; \Omega), \tag{13}$$

$$V(\boldsymbol{r}_{\mathrm{n}1}, \ldots, \boldsymbol{r}_{\mathrm{n}N_{\mathrm{n}}}; \Omega) = \langle \Psi^0_{\mathrm{e(n)}} | H - \hat{T}_{\mathrm{n}} | \Psi^0_{\mathrm{e(n)}} \rangle, \tag{14}$$

which is a many-body potential energy function. This result is not different from what is expected by application of the same argument to a few-body problem, say a molecule. The electron problem appears now in terms of the more familiar language of *bonds* which an appropriate resolution of $V(\{\boldsymbol{r}_{\mathrm{n}j}\}; \Omega)$ defines.

In confronting the interconnected issues of bonding and structure the assumption is usually made that $V(\{\boldsymbol{r}_{\mathrm{n}j}\}; \Omega)$ can be developed in a form

$$V(\{\boldsymbol{r}_{\mathrm{n}j}\}; \Omega) = f(\Omega) + \tfrac{1}{2}\sum_{i,j}{}' \phi^{(2)}_{ij}(\Omega) + \frac{1}{3!}\sum_{i,j,k}{}' \phi^{(3)}_{ijk}(\Omega) + \ldots, \tag{15}$$

where $f(\Omega)$ is a function of volume only (see below) and where the pair, ($\phi^{(2)}$), triplet ($\phi^{(3)}$)..., potentials are assumed to be transferable between configurations, as discussed earlier. This assumption is far from obvious, and in fact is probably not generally correct. So far as the electrons are concerned, there can be no reason to suppose that in the treatment of the statistical mechanics of the ion or nuclear system, configuration independence of the $\phi^{(n)}$ is guaranteed. The existence of transient bonding in metallic liquid states of the traditional semiconductors Si and Ge is already evidence of this (Ashcroft 1990; Štich *et al.* 1989). This point is of some importance to exercises where attempts are made to invert structural information in order to extract forms of $\phi^{(n)}$, especially $\phi^{(2)}$.

For many metals, and also their alloys, it is common practice to assume that the dynamical units of primary statistical concern are single ions, and that associated with those units are transferable potentials $\phi^{(n)}$. Among the simplest of such systems are the sp metals typified by the light alkalis, Mg, Al, and so on. For these it is also usually assumed that the ions are electronically rigid, that is, that the site fluctuations specifically retained in (11 *a*) and (11 *c*) are either ignorable or treatable in a mean-field way. The validity of this assumption will be taken up below, but for the present it is sufficient to note that the approximation being made is rendered by the statement

$$\hat{\rho}_j^{(1)}(\boldsymbol{q}) \rightarrow \langle \hat{\rho}_j^{(1)}(\boldsymbol{q})\rangle = \rho_j(q) = \exp(\mathrm{i}\boldsymbol{q}\cdot\boldsymbol{r}_{\mathrm{n}j}\, g(q)\, Z, \tag{16}$$

where $g(q)$ becomes a *static* form-factor for the ion in question (at high $q, g(q) \rightarrow 1$). Within the same approximation $\hat{h}_j$ is replaced by E_j, a constant, which can be removed by redefinition of the energy origin. Thus by ignoring electromagnetic fluctuation in the ions the hamiltonian now becomes

$$\bar{H} = \hat{T}_{\mathrm{n}} + \frac{1}{2\Omega}\sum_{\boldsymbol{q}} (\hat{\rho}_{\mathrm{n}}^{(1)}(\boldsymbol{q})\,\hat{\rho}_{\mathrm{n}}^{(1)}(-\boldsymbol{q}) - N_{\mathrm{n}})\, Z^2 \bar{v}(q) \tag{17a}$$

$$+\hat{T}_{\mathrm{v}} + \frac{1}{2\Omega}\sum_{\boldsymbol{q}} (\hat{\rho}_{\mathrm{v}}^{(1)}(\boldsymbol{q})\,\hat{\rho}_{\mathrm{v}}^{(1)}(-\boldsymbol{q}) - N_{\mathrm{v}})\, v_{\mathrm{c}}(q) \tag{17b}$$

$$-\frac{1}{\Omega}\sum_{\boldsymbol{q}} \hat{\rho}_{\mathrm{v}}^{(1)}(\boldsymbol{q})\,\hat{\rho}_{\mathrm{n}}^{(1)}(-\boldsymbol{q})\,\omega(q), \tag{17c}$$

where the definition of $\bar{v}(q)$ is now extended to include $g(q)$ as well. This form of hamiltonian is used both in simulation studies (for example, in the method of Car & Parinello (1985)) and also as the starting point for the formal development of multicentre interactions $\phi^{(\mathrm{n})}$ in the metallic state. The latter proceeds (Ashcroft & Stroud 1978) by separating the $q = 0$ terms in (17), which are always large, introducing a coupling constant in (17 *c*), and introducing the *m*th order static response functions $\chi^{(m)}(\boldsymbol{q}_1, \ldots, \boldsymbol{q}_m; \Omega)$ of the homogeneous electron gas hamiltonian (essentially (17 *b*) with $q = 0$ terms removed). Then (Ashcroft & Langreth 1967; Brovman & Kagan 1970)

$$H_{\mathrm{eff}} = \hat{T}_{\mathrm{n}} + \frac{1}{2\Omega}\sum (\hat{\rho}_{\mathrm{n}}^{(1)}(\boldsymbol{q})\,\hat{\rho}_{\mathrm{n}}^{(1)}(-\boldsymbol{q}) - N_{\mathrm{n}})\, Z^2 \bar{v}(q)$$

$$+E_0(\bar{\rho}_{\mathrm{e}}^{(1)})\frac{1}{\Omega}\sum_{m\boldsymbol{q}}\ \sum_{\boldsymbol{q}_1,\ldots\boldsymbol{q}_m} \frac{1}{m+1}\chi^{(m)}(\boldsymbol{q}_1,\ldots,\boldsymbol{q}_m)\, w(q)\, w(q_1)\ldots w(q_m)$$

$$\times\, \hat{\rho}_n^{(1)}(\boldsymbol{q})\,\hat{\rho}_n^{(1)}(\boldsymbol{q}_1)\ldots\hat{\rho}_n^{(1)}(\boldsymbol{q}_m)\,\delta(\boldsymbol{q}+\boldsymbol{q}_1 \mathrm{u}\, \boldsymbol{q}_m), \tag{18}$$

where $E_0(\bar{\rho}_e^{(1)})$ is the ground state energy of the interacting electron gas at the prescribed density $\bar{\rho}_{\mathrm{e}}^{(1)} = N_{\mathrm{v}}/\Omega$ together with the $q = 0$ residues. Note that the static response functions also depend on this density so that for the assumed metallic state, the form of (18) is just a slight generalization of (13), namely

$$H_{\mathrm{eff}} = \hat{T}_{\mathrm{n}} + f(\Omega) + \bar{V}(\boldsymbol{r}_{\mathrm{n}1}, \ldots, \boldsymbol{r}_{\mathrm{n}N_{\mathrm{n}}}; \Omega), \tag{19}$$

where $\bar{V}$ is a state dependent many-centre potential defined via (18), and $f(\Omega)$ is a function solely dependent on density and not-derivable from sums over *n*-atom potentials. In the mean-field view, structural distinctions are made on the basis of the corresponding $\bar{V}$, a many-atom quantity. However, since a system of crystalline symmetry has been developed from a translationally invariant problem, the Fermi surface may not necessarily be given correctly. Correspondingly, the expansion (18) is then not necessarily analytic, as is known from direct summation of energies in two-band and related models. Non-analyticity is also a feature of the tight-binding approach (for example, in the dependence of energetics on coordination number).

The series represented by (18), from which pair and multicentre interactions can now be defined (V. Heine, this symposium), may be termed the mean-field result in view of its neglect of fluctuations. It has been discussed in a detailed review by Hafner (1987) and is also treated extensively in this meeting. It is sufficient to note that the pair term, also intrinsically state dependent, already has the form

$$\phi^{(2)}(r,\Omega)=\frac{Z^2e^2}{r}\left\{1-\frac{2k_{\mathrm{TF}}^2}{\pi}\int_0^{\infty}\mathrm{d}q\,\frac{\sin q}{q^3}\,\frac{\bar{f}(q)}{\epsilon(q,\Omega)}\left(\frac{w(q)}{Zv_{\mathrm{c}}(q)}\right)^2\right\}, \tag{20}$$

where the requisite linear response function has been written

$$\chi^{(1)}(q)=(-k_{\mathrm{TF}}^2/4\pi e^2)\bar{f}(q),$$

where k_{TF} is the Thomas–Fermi wave-vector, ϵ the static dielectric function of the electron gas, and $\bar{f}(q)$ contains all corrections (for exchange and local fields) beyond this. It is clear that at ionic separations typical of a metal, the first term in (20) leads to energies in the range of rydbergs. Yet calculated values of $\phi^{(2)}(r;\Omega)$ are in the millirydberg range, establishing thereby that at the level of *linear* response, only, the second term in (20) cancels the first to within parts per 10^3. For this reason alone contributions to structure and bonding from higher response both static and fluctuating are important to consider. The millirydberg scale of the pair potential represented by (20) provides a useful base for comparison of fluctuation based interactions.

5. Fluctuation-based interactions

The extent to which the mean field approximation implicit in assumption of an electronically rigid ion core constitutes a meaningful approximation is ultimately controlled by the magnitude of the corresponding ion polarizability $\alpha(\omega)$ in a specifically metallic environment. It is known that the polarizability of a free ion differs from that of an ion placed in a metallic environment, but unless conditions are chosen close to the onset of a Mott transition these changes are not large (Nieminen & Puska 1982). The values themselves can, however, be substantial ($\alpha(0)=5.7a_0^3$ for K^+, for example) and the assumption of electronically rigid cores, even for some relatively common systems is therefore often difficult to justify, *a priori*. It should be remembered that a process of local polarization represents a physical mechanism in which by fluctuation charge is separated in space. Since relative displacement is involved, the process is not local and thus not representable by local approximations to true energy functionals. The point being made here is that for fixed $\hat{\rho}_{\mathrm{n}}^{(1)}$ or $\hat{\rho}_{\mathrm{nn}'}^{(2)}$, the electronic problem defined by (11*b*) and (11*c*), and subsequently reduced to (17*b*) and (17*c*) is also the starting point for density functional methods. Density functional theory and its applications has been reviewed recently by Jones & Gunnarsson (1989) (see also Srivastava & Weaire 1987). Since information on exact functionals for inhomogeneous systems is lacking, approximations are usually developed that utilize the properties of the homogeneous counterpart, mainly, however, through local approximation. This approach can only include in an average way the polarization effects that are being addressed here. Similar considerations can clearly apply when charge is displaced by the polarization processes that accompany the excitations of electrons across the gap of a semiconductor.

For cohesion of the condensed state, in particular, the role of electromagnetic fluctuational effects in energetics, has been treated in depth by Barash & Ginzburg (1989). Here the focus is on the more specific problem of the detailed physical form of the microscopic interactions contributing not only to cohesion, but more especially to bonding. The most familiar case occurs when the system is taken to consist of entirely localized charge ($Z = 0$ in (8)) for which (11*a*) is then the only surviving term in the hamiltonian. Then as first shown by London (1930) the pairwise contribution that results is given at long range by

$$\phi_{VdW}(r) = -\hbar \int_0^\infty \frac{d\omega}{2\pi} \alpha_1(i\omega)\,\alpha_2(i\omega) \left\{ \left(\frac{\partial^2 v_c(r)}{\partial r^2}\right)^2 + \frac{2}{r^2}\left(\frac{\partial v_c(r)}{\partial r}\right)^2 \right\} \tag{21}$$

(see also Slater & Kirkwood 1937). Here r is the separation of the two localized systems whose frequency dependent polarizabilities are $\alpha_1(\omega)$ and $\alpha_2(\omega)$. An intuitive and simple limit of (21) results when the polarizability is dominated by a single excitation energy Δ. For identical ions the result is approximately

$$\phi_{VdW}(r) = -\tfrac{3}{4}\Delta(\alpha^{\frac{1}{3}}/r)^6, \tag{22a}$$

where α is the static limit of $\alpha(\omega)$. This form for ϕ_{VdW} can be used to establish the energy scales for fluctuation based interactions in relation to their mean field counterparts whose densities are fixed in the main by $f(\Omega)$ (see (17)). Evidently if Δ is a few rydbergs, r a few a_0, and α for an ion a few a_0^3, then from (22*a*)

$$\phi_{VdW} \sim O(10^{-3})\,\text{Ryd}. \tag{22b}$$

Detailed comparison with (20) requires VdW specification if material dependent parameters (pseudopotentials, densities, and so forth), but it is nevertheless apparent that the fluctuational and mean-field scales can be comparable. This has also been noted in an entirely classical context. In the example of ionic systems, Woodward *et al.* (1988) have shown that thermal fluctuations in mean-field response charge in counter ions assembled around macro-ions can lead to very significant dipole–dipole attraction.

The argument just given does not change in an essential way of the neutral atoms are now replaced by a neutral canonical ensemble of ions and free charge, once again represented by the full hamiltonian (11). Polarizabilities are generally smaller in positive ions than they are in the corresponding neutral atoms, but can be very significant, especially if the ion possesses a filled d-shell. From the expansion of the site charge (8), it is clear from the interaction (11*c*) that a monopole will be coupled to charge fluctuations on neighbouring ions. The ensuing attraction is effectively a two-particle limit of a more general three-particle interaction (see below). If the external monopole were chosen to be an additional electron the interaction would be considered part of the electron affinity (or electronegativity). At long range the attraction, for a monopole $+Ze$, has the well-known form

$$\phi_{md}(r) \sim -(Ze^2/r)\,(\alpha^{\frac{1}{3}}/r)^3, \tag{23}$$

which has a scale similar to (22*a*); like (21) it must be terminated at short range by exchange repulsion. In highly symmetric structures the corresponding energies tend to cancel as can be seen from their formal origin as three-body interactions (Mon *et al.* 1979). For systems lacking such symmetry (quasi-crystalline states, or metallic glasses, for instance) the contributions from (23) can also be significant compared with the standard mean-field result (20).

In considering ions and free charge, rather than neutral atoms, terms (11*b*) and (11*c*) in the fluctuation hamiltonian lead immediately to additional modifications to the conventional lowest-order dispersion results (21) and (22). Let $\epsilon(\boldsymbol{q}, \omega)$ be the frequency dependent dielectric function of the uniform interaction electron gas. Then by including the screening of localized fluctuations (21) is replaced by

$$\phi_{v\,\mathrm{dW}}(r) = \int_0^\infty \frac{\mathrm{d}\omega}{2\pi} \int \frac{\mathrm{d}\boldsymbol{q}}{(2\pi)^3} \int \frac{\mathrm{d}\boldsymbol{q}'}{(2\pi)^3} \alpha^2(\mathrm{i}\omega)\, v_{\mathrm{sc}}(q, \mathrm{i}\omega)\, v_{\mathrm{sc}}(q'; \mathrm{i}\omega) \exp\left[\mathrm{i}(\boldsymbol{q}+\boldsymbol{q}')\cdot r\right] (\boldsymbol{q}\cdot\boldsymbol{q}')^2 \tag{24a}$$

(Mahanty & Taylor 1978; Mon *et al.* 1979; Maggs & Ashcroft 1987). Here $v_{\mathrm{sc}}(\boldsymbol{q}, \omega)$ is the screened Coulomb interaction

$$v_{\mathrm{sc}}(q, \omega) = v_{\mathrm{c}}(q)/\epsilon(q, \omega).$$

Once again, if core-fluctuations are dominated by a single frequency $\varDelta/\hbar$, and if a simple hydrodynamic form for $\epsilon(q, \omega)$ is used, then the van der Waals interaction in a metallic environment is characterized by a scale

$$\phi_{v\,\mathrm{dW}}(r) = -\tfrac{3}{4}\varDelta(\alpha^2(0)/r^6)\,(\varDelta/(\varDelta+\hbar\omega_{\mathrm{p}}))^3, \tag{24b}$$

where ω_{p} is the plasma frequency of the three-dimensional electron gas (and hence $1/\omega_{\mathrm{p}}$ the corresponding characteristic timescale for adjustment of its microscopic fields). When $\varDelta/\hbar > \omega_{\mathrm{p}}$ it is physically obvious that the electron gas can respond only weakly to fluctuations in the core; the expected limit, (22*a*), is therefore readily obtained from (24*b*). It is clear that for realistic choices of α, $\varDelta$ and $\hbar\omega_{\mathrm{p}}$, the scale of (24*b*) may not be very different from (22*a*).

It is worth noting that in addition to the pairwise contributions (24) which has structural significance, there are also fluctuation contributions of a structure independent character. One example is the coupling of the zero-point motion of the multipole operators of the localized charge, to the plasma oscillations of the itinerant charge. If the lincar response function of the latter is $\chi(q, \omega)$ then the contribution corresponding to dipolar coupling is

$$f_{\mathrm{ed}}(\varOmega) = 4\pi \int_0^\infty \frac{\mathrm{d}\boldsymbol{q}}{(2\pi)^3} \chi(q, \mathrm{i}\omega)\, \alpha(\mathrm{i}\omega)\, v_{\mathrm{c}}(q). \tag{25}$$

As shown by Maggs & Ashcroft (1987) this can be a very substantial energy, typically above an electronvolt per electron for reasonable values of α. It quite clearly owes its origin to electronic correlation: in the mean-field sense of band theory it is partly included in the cohesive energy through the self-consistent construction of the one-electron potential. Finally, with respect to the form of valence electron coupling, (11*c*), some especially interesting comparisons can be made between displacive and fluctuational polarization for the case of crystalline order. These bear significantly on the question of electron ordering and are taken up briefly in the appendix.

6. Fluctuational interactions and nonlinear response

The dominant contribution to $\alpha(\omega)$ in an isolated atom is attributable to its outer valence electrons, i.e. to those that are least bound. Suppose that with others this atom then forms a condensed state that is found to be metallic. The valence electron charge, previously localized, has become itinerant; this component of the total electron density was designated earlier as $\langle\hat{\rho}_{\mathrm{v}}^{(1)}(\boldsymbol{r})\rangle$. In a metallic context it is conventionally referred to as screening charge, even though it is recognized that close to the nucleus its form is still reasonably atomic in character. Interstitially it departs from atomic form; nevertheless exactly the same question can be asked about the dynamic or fluctuational structure of $\rho_{\mathrm{v}}^{(1)}(\boldsymbol{r})$ that led in the earlier atomic context to van der Waals attraction.

Consider, first a single ion immersed in an otherwise uniform interacting electron gas. A static distribution $\rho_{\mathrm{v}}^{(1)}(\boldsymbol{r})$ is established around the ion by the arguments given above; it is spherically symmetric. The embedding electron gas is characterized by a response time ω_p – the equivalent of $\hbar/\Delta$ for the atomic case – and on timescales shorter than this, there will be microfields relative to a positive background with a full set of multipole components.

Suppose that a second ion is now introduced; it establishes its own screening charge, and once again these will also have a multipole decomposition on timescales short compared with ω_{p}. These observations are already sufficient to show that the arguments given for the atomic case, where the outer valence charge was considered localized but fluctuating, can be repeated for the case of screening charge. The viewpoint is actually very little changed; fluctuating multipole attraction is now identified with certain higher-order response terms. In fact, Maggs & Ashcroft (1987) have referred to this picture as one involving the fluctuation of pseudo-atoms, thereby extending into the dynamic régime the earlier static pseudo-atom concept of Ziman (1967). The viewpoint is an especially useful one in the context of metal–insulator transitions of the band-overlap type since the fluctuating dipole interaction is the embodiment of fluctuation physics that is common on both sides of the phase boundary (Ashcroft 1990; Goldstein *et al.* 1989).

The simplest application of this picture arises when a simple monopole (a rigid ion, or an electron) is found in the neighbourhood of a fluctuating pseudo atom. For this situation the corresponding contribution to a pair interaction has actually been discussed already, for it is nothing but the two-site limit of those terms in the response sequence (18) that are beyond linear. With this observation the source of attraction underlying the chemical concepts of electron affinity or electronegativity is also identified. It may be noted in passing that a qualitative association between the interaction of an electron with higher fluctuation processes and the occurrence of superconductive ordering has already been recorded by Luo & Wang (1987) (in the language of electron affinity) and by Ichikawa (1989) (in terms of electronegativity).

More interesting is the case where fluctuations on two sites are coupled; here a direct equivalence to (21) or (23) emerges except that the fluctuations originate with screening charge which in more common approximations is regarded as static. In the three dimensional examples so far considered, the corresponding contribution to pairwise interactions is

$$\phi_{\mathrm{ff}}(r)=\int_0^{\infty}\frac{\mathrm{d}\omega}{2\pi}\int\frac{\mathrm{d}\boldsymbol{q}}{(2\pi)^3}\int\frac{\mathrm{d}\boldsymbol{q}'}{(2\pi)^3}v_{\mathrm{sc}}(\boldsymbol{q}+\boldsymbol{q}',\mathrm{i}\omega)\,v_{\mathrm{sc}}(\boldsymbol{q}',-\mathrm{i}\omega)\,\omega_{\mathrm{s}}(\boldsymbol{q})$$
$$\times[\Lambda^{(3)}(\boldsymbol{q}+\boldsymbol{q}',\mathrm{i}\omega;\boldsymbol{q}',-\mathrm{i}\omega;\boldsymbol{q},0)]^2\,\mathrm{e}^{\mathrm{i}\boldsymbol{q}\cdot\boldsymbol{r}},\quad(26)$$

where $\omega_{\mathrm{s}}(q)$ is now a statically screened pseudopotential. In (26) $\Lambda^{(3)}$ is the irreducible three-point function of the electron gas (Cenni & Saracco 1988). For small $\boldsymbol{q}$ and $\boldsymbol{q}'$ it behaves as $(1/\omega^2)\,\boldsymbol{q}\cdot\boldsymbol{q}'$ which may be compared with $(\alpha(0)/(\omega^2-\Delta^2))\,\boldsymbol{q}\cdot\boldsymbol{q}'$ as appropriate in the simplest approximation to the localized case. The analytic behaviour of $\phi_{\mathrm{ff}}(r)$ is again a power-law attraction at long range (*ca.* $-1/r^6$ in three dimensions). Dimensional analysis refines this power law to

$$\phi_{\mathrm{ff}}(r) \sim -Z^2(r_{\mathrm{s}}^{\frac{9}{2}}/r^6)\,h(r_{\mathrm{s}}) \tag{27}$$

in atomic units. Here $h(r_{\mathrm{s}})$ is any dimensionless function of density ($\frac{4}{3}\pi r_{\mathrm{s}}^3 a_0^3 = 1/\rho_{\mathrm{e}}^{(1)}$) and is typically of order 10^{-2} (Maggs & Ashcroft 1987). The possible numerical significance of higher-order dynamic processes has been noted by Rasolt & Geldart (1975) and also by Langreth & Vosko (1987) who also show that higher-order terms modify (27) still further but only to the extent of a logarithm. The magnitude of the fluctuation attraction (relative to mean field) is traceable to the fact that static screening is always constrained by the perfect screen sum rule ($\lim_{q\to 0}\epsilon^{-1}(q,0)\to 0$). This constraint does not hold at finite frequencies so that static interactions which involve intermediate higher-order processes can be stronger than their formal order might suggest. The major physical consequence is that in addition to mean-field results typified by (20), corrections arise from electromagnetic fluctuation in both localized charge (see (23)) and in response charge (see (26)). These can be expected to have structural significance according to circumstance, even though they may contribute little to overall cohesion in a metallic state.

7. Fluctuational attraction in the homogeneous and inhomogeneous electron gas

An attractive interaction, entirely equivalent to (26) also exists between electrons in the three dimensional electron gas, as shown also by Maggs and Ashcroft (1987). The underlying physical origin is also equivalent: around any electron is formed a static distribution of response or correlation charge $\langle\hat{\rho}_{\mathrm{ee}}^{(2)}(o,\boldsymbol{r})\rangle = \rho_{\mathrm{e}}^{(1)}g(r)$, where $g(r)$ is the pair correlation function. If the electron and surrounding correlation charge is then probed by another electron whose function is to serve as a test charge, the standard static screened interaction results and it is this combination that is controlled by the perfect screening sum rule. However, if the test electron is also accompanied by its own screening charge then fluctuations in both screening charges (or correlation shells) lead by the arguments in .§.6 to the long-range attraction

$$\phi_{\mathrm{eff}}(r) = \int_0^\infty \frac{\mathrm{d}\omega}{2\pi}\int\frac{\mathrm{d}\boldsymbol{q}}{(2\pi)^3}\int\frac{\mathrm{d}\boldsymbol{q}'}{(2\pi)^3}\,v_{\mathrm{sc}}(\boldsymbol{q}+\boldsymbol{q}',\mathrm{i}\omega)\,v_{\mathrm{sc}}(\boldsymbol{q}',-\mathrm{i}\omega)\,v_{\mathrm{sc}}(\boldsymbol{q},0) \times [\Lambda^{(3)}(\boldsymbol{q}+\boldsymbol{q}',\mathrm{i}\omega;\boldsymbol{q}',-\mathrm{i}\omega;\boldsymbol{q},0)]^2\,\mathrm{e}^{\mathrm{i}\boldsymbol{q}\cdot\boldsymbol{r}} \tag{28}$$

and once again dimensional analysis in combination with the resulting r^{-6} power law give

$$\phi_{\mathrm{eff}}(r) \approx -(r_{\mathrm{s}}^{\frac{9}{2}}/r_{\mathrm{s}}^6)\,h(r_{\mathrm{s}}).$$

This term, which is significant in magnitude, goes considerably beyond what is expected on the basis of random-phase or related approximations to the local field problem in the interacting electron gas. Power law decay is clearly of a different analytic character from either the exponentially decaying (Thomas–Fermi) or Friedel oscillatory behaviour (Lindhard) which typify approximations to effective electron–electron interactions that normally ignore fluctuations. Its origin lies with physical processes that cannot be correctly included in the standard local applications of homogeneous electron gas results to inhomogeneous problems. Again, this is because polarization is fundamentally a process involving the relative displacement of charge, and as such is not local

Given this observation, and the product structure of the kernel in (28) an approximation that is useful in the inhomogeneous context can, however, be made. This is merely to associate separate three-point functions with each of two local densities characteristic of a pair of points in a non-uniform electron gas. Rapcewicz & Ashcroft (1990) have shown that when this procedure is applied to two isolated atoms then van der Waals attraction between the two centres is recovered, and it possesses essentially the correct magnitude. Incorporation of fluctuation therefore leads to a significant correction to the results obtained by Gordon & Kim (1972) in their study of attraction between closed shell atoms using functional methods but in a strictly local approximation. This corroborates the general point being made that functional approaches that omit fluctuations, represented here by only the simplest nonlinear corrections to response, can err qualitatively in their predictions of bonding.

Appendix A. Fluctuation and polarization in crystalline states

For crystalline states nuclear motion is usually described as an N_n-particle small oscillation problem in the harmonic or self-consistent harmonic approximation. In a Bravais lattice, the coordinate of an ion is normally written

$$\boldsymbol{r}_{nj} = \boldsymbol{R}_j + \boldsymbol{u}_j,$$

where $\boldsymbol{R}_j = \langle \boldsymbol{r}_{nj} \rangle$ is the equilibrium average of the coordinate, and defines one of the N_n sites of the crystal. For electronically rigid ions the first correction beyond the terms normally defining the band-structure becomes (see (13*c*))

$$\sum_j \sum_q e^{i\boldsymbol{q}\cdot\boldsymbol{R}_j} \hat{\rho}_v^{(1)}(\boldsymbol{q}) (i\boldsymbol{q}\cdot\boldsymbol{u}_j) w(q). \tag{29}$$

However, if fluctuations in the ion are now restored, then (29) is replaced by

$$\sum_j \sum_q e^{i\boldsymbol{q}\cdot\boldsymbol{R}_j} \hat{\rho}_v^{(1)}(\boldsymbol{q})\, i\boldsymbol{q}\cdot(\boldsymbol{u}_j + \hat{d}_j/Ze) \tag{30}$$

to lowest order. This clearly displays the dual sources of polarization namely displacive (*ca.* $\boldsymbol{u}_j$) from phonon polarization waves, and local fluctuation (*ca.* $\hat{d}_j/Ze$). In a crystalline environment the latter also admit of waves of polarization, but in a different frequency range. The combination of the two in (30) shows very clearly the possibility of interference, the more so for lower valent materials ($Z = 1$). In an ionic crystal that is at the same time metallic the form of the electron coupling will differ from (30) in that the monopole parts can substantially cancel (depending on relative sizes of the positive and negative monopole charges). On the other hand, relative to the motion of a cell centre of mass there will also be dipolar character coupling arising from optic phonons in which these charges will now effectively add. Polar-optic coupling can therefore be especially large.

Typical values of $\langle u_j^2 \rangle^{\frac{1}{2}}$ and $\langle (d_j/Ze)^2 \rangle^{\frac{1}{2}}$ to be associated with (27) are

$$\text{(i)} \quad \langle u_j^2 \rangle^{\frac{1}{2}} = 0.24(r_s^3/AZ)^{\frac{1}{4}} a_0,$$

where A is the nuclear mass number (a result which follows from the known moments of the Coulomb harmonic problem), and

$$\text{(ii)}\quad \langle (d_j/Ze)^2\rangle^{\frac{1}{2}} \sim a_0(2(\alpha_0/a_0)^3)\,(\bar{\Delta}(Ry))^{\frac{1}{2}},$$

where $\bar{\Delta}$ is an average excitation energy of the ion whose polarizability is α. It is not difficult, therefore, to envisage situations where these two sources of polarization lead to (*a*) comparable coupling, (*b*) displacive coupling dominating internal fluctuation (e.g. in the sp metals characterized by small α) and (*c*) internal fluctuation coupling dominating displacive (large Z, and in metals with complete or incomplete but polarizable d-shells). In the conventional view of electron pairing transitions (as in the superconductive state) it has been common to regard phonon-polarization waves (*ca.* $\boldsymbol{u}_j$) or the Fröhlich interaction (Fröhlich 1960) as the source of the requisite electron coupling. The possibility that charge fluctuation (*ca.* $\hat{d}_j/Ze$) might possibly dominate was first put forward by Little (1964) and subsequently applied to systems with itinerant and quasi-localized electrons by Geilikman (1965). An extensive discussion of the interplay of both sources of coupling in the context of specific materials, and especially layered compounds, has been given by Ginzburg & Kirzhnits (1982).

For crystalline symmetry, the localized fluctuations can also be coherent, as pointed out by Hopfield (1958), by Anderson (1963), by Lundqvist & Sjölander (1963), and by Lucas (1968). The nature of such collective excitations parallels that of the phonons; they are polarization waves resulting from dipolar coupled fluctuations on each crystalline site. A single-particle band-structure can still be defined for this case: it will arise by excerpting the zero-phonon-terms from the normal mode structure that results from coupled polarization-wave systems. Then the valence electron interaction term going beyond terms specifically included in band-structure will not be (32) but rather

$$\sum_j \sum_{\boldsymbol{q}} \mathrm{e}^{-W(q)} \exp(\mathrm{i}\boldsymbol{q}\cdot\boldsymbol{R}_j)\,\hat{\rho}_{\mathrm{v}}^{(1)}(\boldsymbol{q})\,\mathrm{i}\boldsymbol{q}\cdot(\boldsymbol{u}_j+\hat{d}_j/Ze), \tag{33}$$

where $W(q)$ is a Debye–Waller factor corresponding to the zero-phonon contributions (Ashcroft 1989). Dynamics of both ion motion and internal charge then enter the coupling implicitly and it is evident that inferences made in such cases from, say, the isotope effect in superconductivity, and that are based on $\boldsymbol{u}_j$ alone can hardly be complete.

In three dimensions, the fluctuational polarization-wave structure is typified by an energy scale Δ, and by a dispersion $\hbar(\rho_{\mathrm{n}}\, e^2/m)^{\frac{1}{2}}$. In contrast to this, the ordinary plasmon in two dimensions has no energy gap at $q=0$. Accordingly one may expect the coupled polarization wave system to disperse to $q=0$ in a two-dimensional crystalline metal with localized fluctuating charge. Such modes have the capability of accepting arbitrarily small energies, as do acoustic phonons in three dimensions, which should be of some significance in low-temperature electronic transport. Though the collective dynamics of the two-dimensional electron gas are quite different from the three-dimensional case, the arguments leading to fluctuation based electronic attraction still carry through. It is general consequence of reduced dimensionality that the role of fluctuations increases relative to that of mean field. Thus, as Rapcewicz & Ashcroft (1990) show, the equivalent of (30) leads to attraction with long-ranged behaviour $-r_{\mathrm{s}}^3/r^{\frac{9}{2}}$ (compared with $-r_{\mathrm{s}}^{\frac{9}{2}}/r^6$ in three dimensions). The general arguments of Kohn & Luttinger (1965) then lead to the expectation of unstable behaviour, most notably of a pairing character.

References

Anderson, P. W. 1963 *Concepts in solids*, p. 132ff. New York: Benjamin.

Ashcroft, N. W. 1981 Ordered ground states of metallic hydrogen and deuterium. In *Physics of solids under high pressure* (ed. J. S. Schilling & R. N. Shelton), pp. 155–160. Amsterdam: North-Holland.

Ashcroft, N. W. 1989 Quantum solid behavior and the electronic structure of the light alkali metals. *Phys. Rev.* B **39**, 10552–10559.

Ashcroft, N. W. 1990 Electronic fluctuation, the nature of interactions and the structure of liquid metals. *Nuovo Cim.* D **12**, N4–5; 597–618.

Ashcroft, N. W. & Langreth, D. 1967 Compressibility and binding energy of simple metals. *Phys. Rev.* **155**, 682–684.

Ashcroft, N. W. & Stroud, D. 1978 Theory of the thermodynamics of simple liquid metals. *Solid St. Phys.* **33**, 1–81.

Barash, Yu. S. & Ginzburg, V. L. 1989 Electromagnetic fluctuations and molecular forces in condensed matter. In *The dielectric function of condensed systems* (ed. L. V. Keldysh, D. A. Kirzhnits & A. A. Maradudin) pp. 389–457. Elsevier.

Brovman, E. G. & Kagan, Yu. 1970 Long wavelength phonons in metals. *Soviet Phys. JETP* **30**, 721–727.

Car, R. & Parrinello, M. 1985 Unified approach for molecular dynamics and density-functional theory. *Phys. Rev. Lett.* **55**, 2471–2474.

Cenni, R. & Saracco, P. 1988 Evaluation of a class of diagrams useful in many-body calculations. *Nucl. Phys.* A **487**, 279–300.

Chihara, J. 1985 Liquid metals and plasmas as nucleus–electron mixtures. *J. Phys.* C **18**, 3103–3118.

Fröhlich, H. 1960 The theory of the superconductive state. *Rep. Prog. Phys.* **24**, 1–23.

Geilikman, B. T. 1965 A possible mechanisms for superconductivity in alloys. *Soviet Phys. JETP* **21**, 796–798.

Ginzburg, V. L. & Kirzhnits, D. A. (eds) 1982 *High temperature superconductivity.* New York: Consultants Bureau.

Goldstein, R. E., Parola, A. & Smith, A. 1989 Fluctuating pseudoatoms in metallic fluids. *J. chem. Phys.* **91**, 1843–1854.

Gordon, R. G. & Kim, Y. S. 1972 Theory for the forces between closed-shell atoms and molecules. *J. chem. Phys.* **56**, 3122–3133.

Hafner, J. 1987 *From hamiltonians to phase diagrams.* Berlin: Springer-Verlag.

Heine, V. 1970 The pseudopotential concept. *Solid St. Phys.* **24**.

Hopfield, J. 1958 Theory of the contribution of excitons to the complex dielectric constant of crystals. *Phys. Rev.* **112**, 1555–1567.

Ichikawa, S. 1989 Superconductivity and optimum electronegativity. *J. Phys. Chem. Solids* **50**, 931–934.

Jones, R. O. & Gunnarsson, O. 1989 The density functional formalism, its application and prospects. *Rev. mod. Phys.* **61**, 689–745.

Kohn, W. & Luttinger, J. M. 1965 New mechanism for superconductivity. *Phys. Rev. Lett.* **15**, 524–526.

Langreth, D. & Vosko, S. H. 1987 Exact electron-gas response functions at high density. *Phys. Rev. Lett.* **59**, 497–500.

Little, W. 1964 Possibility of synthesising an organic superconductor, *Phys. Rev.* A **134**, A 1416–1424.

London, F. 1930 Zur theorie und systematik der Molekularkräfte. *Z. Phys.* **63**, 245–279.

Lucas, A. 1968 Collective contributions to the long range dipolar interaction in rare-gas crystals. *Physica* **35**, 353–368.

Lundqvist, S. & Sjölander, A. 1963 On polarization waves in van der Waals crystals. *Arkiv. Fysik* **26**, 17–34.

Luo, Q. & Wang, R. 1987 Electronegativity and superconductivity. *J. Phys. Chem. Solids* **48**, 425–430.

Maggs, A. C. & Ashcroft, N. W. 1987 Electronic fluctuation and cohesion in metals. *Phys. Rev. Lett.* **59**, 113–116.

Mahanty, J. & Taylor, R. 1978 Van der Waals forces in metals. *Phys. Rev.* **17**, 554–559.

Moody, J., Shapere, A. & Wilczek, F. 1989 Adiabatic effective lagrangians. In *Geometric phases in physics* (ed. A. Shapere & F. Wilczek), pp. 160–183. Singapore: World Scientific.

Mon, K. K., Ashcroft, N. W. & Chester, G. V. 1979 Core polarization, and the structure of simple metals. *Phys. Rev.* B **19**, 5103–5122.

Moriarty, J. 1990 Analytic representation of multi-ion interatomic potentials in transition metals. *Phys. Rev.* B **42**, 1609–1628.

Nieminen, R. & Puska, M. J. 1982 Core polarizabilities in metals. *Physica Scr.* **25**, 952–956.

Rapcewicz, C. & Ashcroft, N. W. 1990 Fluctuation attraction in condensed matter: a non-local functional approach. (Submitted.)

Rasolt, M. & Geldart, D. J. W. 1975 Gradient corrections in the exchange and correlation energy of an inhomogeneous electron gas. *Phys. Rev. Lett.* **35**, 1234–1237.

Rehr, J. J., Zaremba, E. & Kohn, W. 1975 Van der Waals forces in the noble metals. *Phys. Rev.* B **12**, 2062–2066.

Roberts, S. 1950 A theory of dielectric polarization in alkali-halide crystals. *Phys. Rev.* **77**, 258–263.

Slater, J. C. & Kirkwood, J. G. 1937 The van der Waals forces in gases. *Phys. Rev.* **37**, 682–697.

Srivastava, G. P. & Weaire, D. 1987 The theory of the cohesive energy of solids. *Adv. Phys.* **36**, 463–517.

Štich, I., Car, R. & Parrinello, M. 1990 Bonding and disorder in liquid silicon. *Phys. Rev. Lett.* **63**, 2240–2243.

Sugiyama, G., Serah, G. & Alder, B. J. 1989 Ground-state properties of metallic lithium. *Physica* A **156**, 144–168.

Woodward, C. E., Jönsson, B. & Åkesson, T. 1988 The ionic correlation contribution to the free energy of spherical double layers. *J. chem. Phys.* **89**, 5145–5152.

Ziman, J. M. 1967 Some non-structural aspects of the theory of metals. *Proc. phys. Soc.* **91**, 701–723.

Discussion

L. J. Sham (*University of California, San Diego, U.S.A.*). In the 1960s, Luttinger & Kohn found an interaction term between electrons in the homogeneous electron gas which cause $l > 0$ Cooper pairing and, thus, a superconducting instability. What is the relation between this interaction and the fluctuation term which Professor Ashcroft discussed?

N. W. Ashcroft. The term I was primarily discussing was the first of a ladder sequence and corresponds to the first of the multipole terms that result when fluctuations in normal static response is explicitly excepted from the totality of nonlinear response terms. Kohn & Luttinger do indeed include nonlinear interactions and one in particular is just the exchange modification of the fluctuating dipole–dipole contribution developed in my paper. On length scales important for the formation of a Cooper pair, this latter contribution is significantly attractive. However, to determine the likely angular momentum state it is necessary to know the form of the effective interaction at shorter length scales. This has not been settled in detail, though it is clear on general grounds that for $r < 2r_s$ (where the response charge around each of a pair of electrons begins to overlap) the attractive character must start to wash out.

P. W. ANDERSON (*Princeton, U.S.A.*). (1) Is this effect to some extent responsible for the high stability of certain heavy ions, so noticeable in oxide systems: such as Ba^{2+}, La^{2+}, Bi^{3+}? (2) With regard to long-range electron–electron forces, how can one justify using the lowest-order polarization bubble at very low density when the series is considered generally to be a series in powers of r_s? Incidentally, my paper shows that perturbation theory in general is in great difficulty in two dimensions in quite a deep sense.

N. W. ASHCROFT. (1) The formation (and implied stability) of ions in a compound or intermetallic is associated with a charge transfer process. The ionization energy penalty that is paid is normally recouped from two sources, namely, electrostatic Madelung energy (in the state of eventual charge transfer), and electron affinity which impels the process in the first place. It is to electron affinity that my remarks on fluctuational aspect most naturally apply here. An electron placed outside a localized distribution of charge is attracted by an amount that depends on the nature of the localized charge. If the electron is subsequently bound a new distribution results with a new, and larger polarizability. A classic example is O^{2-}.

(2) In fact the diagrams discussed were not of the bubble class, but of the ladder class. The bubble class would lead to a normal static distribution of screening charge around an electron, and as such this would be constrained by the perfect screening sum-rule. The central physics of the effects being discussed here (as illustrated by the first term in the ladder sequence) is that internal dynamic loops are rife. These are not constrained by the rule, and because of this terms which have a formal order higher than bubble equivalents can actually lead to larger effects. While the r_s expansion (in two or three dimensions) might well be questioned for, say, thermodynamic functions, the focus here is on a specific subset of terms namely those of a predominantly inverse power multipole character that contribute to an effective interaction.

B. COLES (*Imperial College, London, U.K.*). Treatment of ion core fluctuations takes the polarizability as a well-defined quantity, but it depends on energy denominators which will be very different for Cu $3d^{10}$ ion or $3d^{10}4s$ atom and the 3d full core in copper metal. Does this make it impossible to use Professor Ashcroft's approach to calculate the full 3d shell contribution to cohesion of metallic copper?

N. W. ASHCROFT. The polarization $\alpha(\omega)$ that enters is the one appropriate to the actual states of the system, here a metal where the electrons have been partitioned into valence and localized. Since both energy denominators and matrix elements of the dipole operator differ (for free ions and ions embedded in a metallic distribution of valence electrons) the values of $\alpha(\omega)$ indeed reflect such differences. The changes are discussed in some detail by Nieminen & Puska. A determination of the van der Waals contribution to ion–ion potentials therefore merely requires a prior determination of *in situ* ionic polarizabilities.

P. B. ALLEN (*SUNY, Stony Brook, New York, U.S.A.*). There are multi-atom fluctuation-induced interactions: the Axilrod–Teller term is the three-body analogue of the van der Waals r^{-6} interaction. What are the relevance and properties of these terms?

N. W. ASHCROFT. Rapcewicz and I have argued that the successful determination of long-range van der Waals attraction between a pair of atoms, starting only with the properties of the homogeneous interacting electron gas, can be taken as supporting evidence for the necessity to go beyond static response in the construction of effective electron–electron interactions. If this view point is correct, then a simple logical extension of the argument also has to be correct namely that a three-atom interaction with the expected form should emerge when a third distribution of localized charge is introduced into the previous pair. This is exactly what Rapcewicz and I do discover: the three-atom interaction that emerges has precisely the Axilrod–Teller form, and once again the magnitude is quite well given, starting from electron gas information alone.

Incommensurate structures

By J. D. C. McConnell
Department of Earth Sciences, Parks Road, Oxford OX1 3PR, U.K.

Incommensurate structures occur in a very wide range of crystalline materials and illustrate a number of interesting aspects of the physics of atomic interaction in the context of crystal structure and bonding. Formal theories that deal with the origin of these phases are reviewed and shown to have a great deal in common. Several important examples of incommensurate structures in simple compounds are described and these include the minerals quartz and nepheline. Nepheline, which has not been analysed previously shows an interaction between oxygen displacements associated with the loss of a triad axis, a three-state Potts model, and potassium-vacancy ordering. The important role of incommensurate structures in mineral solid solutions is discussed and illustrated with reference to the plagioclase feldspar solid solution.

1. Introduction

Incommensurate structures are characterized by the existence of a modulation, or system of modulations, which are incommensurate with the underlying lattice repeats of the crystal. Evidence of such behaviour is generally obtained from single crystal diffraction data where the presence of the modulation is associated with the existence of additional closely paired intensity maxima that are convoluted with either the normal reciprocal lattice points, or with high symmetry points on the Brillouin zone boundary. This distribution of intensity may be specified formally by defining a small reciprocal vector Q as measured from the nearest symmetry point. This vector characterizes the wavelength and wave vector for the modulation, it normally has high symmetry and usually defines a wavelength of the order of several unit cell, or supercell, repeats. To date a large number of incommensurate structures have been studied. Examples include simple compounds such as biphenyl, $NaNO_2$ and quartz but there is also a large class of mineral examples where incommensurate behaviour may be directly associated with ordering behaviour in solid solutions. This article deals initially with the theory of the origin of incommensurate structures in simple compounds and then proceeds to the special problems associated with incommensurate behaviour in solid solutions. Problems in the latter category have not been discussed in a general way previously. Sections 3 and 4 provide examples of incommensurate structures in simple compounds and solid solutions respectively.

2. General theory

In attempting to provide a completely general theory for the origin of incommensurate crystal structures it is necessary to establish the nature of interactions within the single crystal that can lead to a minimum of free energy for some arbitrary wavelength associated with the observed modulation. It is usual in the case of simple ordering events in single crystals to find that positive gradient energy terms associated with modulation (non-zero values of Q) lead to an increase in enthalpy and free energy. Thus it is usual to find energy minima located at the symmetry points where, by definition, Q is zero. A mechanism or interaction which acts to reduce the free energy of the system for non-zero Q may be anticipated where there are competing local interactions, as in the ANNNI model. However, a phenomenological approach based on local neighbour interactions is not particularly helpful in developing a general theory of incommensurate behaviour.

To provide a sufficiently abstract theory it is useful to begin by discussing the formal symmetry aspects of the problem. Symmetry arguments have been used by a number of workers and it is possible to show that all the theories proposed have a great deal in common although this may not be apparent at first sight. The earliest recorded reference to symmetry in the development of incommensurate structures is contained in Axe *et al.* (1970). These authors presented a theory of anomalous acoustic dispersion associated with interaction between two phonon bands of the same reduced symmetry, and comparable frequency, leading to an energy minimum in the lower band at a non-zero value of Q. The same principle was evoked somewhat later (McConnell 1978) in defining the selection rules for the interaction of modulated structures based on ordering schemes of different symmetry. It was noted there that it is possible to chose two ordering schemes of appropriate symmetry for a symmetry point vector such that they have the same reduced symmetry at Q. Thus the ordering structures may interact with advantage at non-zero Q to produce a complex modulation and hence reduce the free energy of the system. It was also noted that, within the complex modulation thus formed, the two modulation components must be in quadrature and hence may be written:

$$\eta \cos Qr \pm \xi \sin Qr, \tag{1}$$

where η and ξ are order parameters associated with the two original ordering schemes (group representations) and the choice of sign relates to the choice of the lower of two free energies. This concept of the interaction of two component structures within the modulation has been particularly useful in establishing the structure of incommensurate phases in insulators (Heine & McConnell 1981). It has led directly to the development of the so-called gradient ploy (Heine & McConnell 1981, 1984) which may be used both to determine the second structure, the nature of the interaction and the relevant phase relationships. In dealing with the component structures in a modulation in this way it is possible to demonstrate the precise symmetry properties which define the gradient concept and it will be convenient to illustrate these later in the context of the incommensurate structures of quartz and nepheline.

An alternative but closely related analysis of the origin of incommensurate phases arose originally from considering the implications of the so-called Lifshitz invariant. This was originally discussed in the context of the selection rules associated with second-order phase transformations. Lifshitz (1942) was the first to point out that the presence of an invariant of the form:

$$\eta \, \mathrm{d}\xi/\mathrm{d}x - \xi \, \mathrm{d}\eta/\mathrm{d}x, \tag{2}$$

in the Landau free energy expansion would lead to a spatial instability in the transforming single crystal, and thus preclude the occurrence of a normal second-order phase transition. Here η and ξ are order parameters associated with two degenerate one-dimensional representations and the spatial derivatives relate to a chosen direction in the crystal. The Lifshitz invariant arises naturally for representations associated with vectors on the Brillouin zone boundary under conditions where the symmetry includes glide or screw operations. Later it was appreciated that the presence of this invariant must lead to the development of a modulated or incommensurate structure (Levanyuk & Sannikov, 1976). It is possible† to prove that the ordering selection rules defined by the Lifshitz gradient invariant are equivalent to those already discussed in the two component structure theory in (1) above, and that the gradient invariant is simply a formal description of their symmetry relationships.

In discussing gradient invariants in general Sannikov & Golovko (1989) distinguishes two possibilities relating to the origin of incommensurate phases. The first of these they identified with a true Lifshitz invariant where the two representations (and related order parameters) were combined in a two-dimensional irreducible representation, and were therefore necessarily degenerate in energy. In the second case, which they describe as a Lifshitz-type invariant, the two representations are initially independent and, consequently, need be merely comparable in energy to interact. This distinction, which we here describe as the Sannikov condition, is important physically for the following reasons. In the case of a true, or what we will here call a proper Lifshitz invariant the basis structures to be considered in the incommensurate structure are necessarily identical physically and must transform into one another under the symmetry operations of the vector associated with the relevant symmetry point. In the case of incommensurate structures associated with the Lifshitz-type invariant, however, the component structures, which are now defined in terms of initially independent one-dimensional irreducible representations, certainly need not transform into one another physically. They may well relate to totally different physical ordering events or to an ordering mode combined with a soft mode.

In the present article I use the term 'improper Lifshitz invariant' to describe this more general situation. Thus we use the term to describe the situation where certain symmetry criteria and phase relationships are satisfied but the structural phenomena to which they relate are not degenerate. Examples of both types of incommensurate structure are described in the literature. In biphenyl, for example, the incommensurate structure designated phase III by Baudour & Sanquer (1983) stable below 20 K, must be associated with the existence of a proper Lifshitz invariant. This follows since the high-temperature space group of biphenyl is $P2_1/a$, and modulation wave vector in phase III is associated with the symmetry point vector $0, \frac{1}{2}, 0$ where all representations are necessarily doubly degenerate through the presence of the screw diad axis. In studying the nature of the incommensurate structure in this case it is only necessary to establish one component structure or structural event since the second is identical and must be phase related within the modulation to the first. There are many examples in the literature of structures associated with improper Lifshitz invariants. Sodium nitrite is an important example in this class and shows

† The operative condition for the Lifshitz invariant is that the anti-symmetrized square of the chosen irreducible two-dimensional representation for the crystal should contain a representation that transforms as a vector. Now for appropriate pairs of full space group representations in the two component structure theory which have the name symmetry at Q the symmetry elements that turn Q into $-Q$ must have opposite sign since the representations are orthogonal. Thus the product of all pairs of characters in the two representations in the subset of elements that turn Q into $-Q$ must yield the character -1. This condition in turn dictates that the product representation is itself a vector representation.

incommensurate behaviour in a narrow temperature interval around 163 °C (McConnell 1981*a*; Heine & McConnell 1981). At temperatures above 163 °C sodium nitrite is paraelectric with space group *Immm*. The incommensurate structure is modulated parallel to the *a* axis and two-component structure theory indicates that one of the structures is ferroelectric with space group *Im2m*, while the other is ferroelastic with space group $I2/m_z$.

The distinction, originally created by Sannikov, between what are here described as proper and improper Lifshitz invariants is very important when one considers the potential stability of incommensurate structures over a range of temperatures. Where the origin of the incommensurate phase is associated with a proper Lifshitz invariant there need be no reason in symmetry to seek a low-temperature commensurate phase and the IC structure, or something very closely related to it, may well exist down to 0 K. However, where the incommensurate structure, at least in a simple compound, is associated with an improper Lifshitz invariant it is usual to find the incommensurate phase restricted to a narrow temperature interval above a first-order transformation to a commensurate structure related to the dominant-component structure. These conclusions accord well the actual behaviour of incommensurate structures in simple compounds such as that in quartz and sodium nitrite. Up to this point the theory of incommensurate behaviour has revolved around the possibilities associated with the proper and improper Lifshitz invariants, and the existence of incommensurate structures that can be analysed on strict symmetry principles.

There is in addition a large and important class of incommensurate structures that occur in binary solid solutions and here the appropriate theory is less well developed. While the two-component theory has been used (McConnell 1978) and is often applicable, it is also clear that the rigidity of the symmetry rules must be substantially relaxed in the interests of variable chemistry. The phenomena are illustrated in the case of the incommensurate structure of wustite, $Fe_{(1-x)}O$ (Garstein *et al.* 1986) and also in one of the most important mineral solid solution systems, namely the plagioclase feldspars. In both these materials the incommensurate mechanism is complicated by the fact that it is necessary to consider ordering characteristics as a function of composition as well as temperature. It is at this point that the simple physical picture associated with the mixing and splitting of phonon bands, which appeared at the beginning of this section, proves to be inadequate.

It has proved helpful in practice to develop a theory of incommensurate structures in binary solid solutions by first establishing possible ordered structures for specific end member phases, or for intermediate compounds, in the relevant binary system. Where the symmetry of these ordered structures satisfy the Lifshitz invariant condition the structures may combine in an incommensurate modulation with appropriate phase and a substantial reduction in free energy. The refractory phase mullite provides a good example of the application of these principles.

Mullite is the most important refractory phase in the system SiO_2–Al_2O_3 and develops an incommensurate structure on crystallization at 1500 °C (McConnell, 1981*a*; McConnell & Heine 1985*b*). In terms of possible end members it can be considered as a member of a solid solution series between end member compounds $Al_8Si_4O_{20}$ (sillimanite) and iota alumina, with the corresponding formula unit written as $Al_{12}O_{18}[\,]_2$, where the symbol [ ] indicates the presence of oxygen vacancies. When disordered both these compounds are isostructural with space group Pbam. On ordering each develops a simple superlattice involving doubling of the *c* axis but with space groups *Pbnm* (ordering of Al and Si as in sillimanite) and *Pnnm* (ordering of vacancies as in iota alumina). It is easy to show that these two space groups are related in an improper Lifshitz invariant for modulation parallel to the *a* crystallographic axis, which accords with the experimental diffraction observations (Cameron 1977). Within the incommensurate structure Si/Al ordering and the ordering of vacancies combine to reduce the free energy of the system. Clearly the real component structures present must depart appreciably from the idealized end member structures for sillimanite and iota alumina.

A comparable approach may be used in the study of the incommensurate structure in the plagioclase feldspar solid solution although, since these materials are triclinic the symmetry aspects are by no means so clear cut. The details will be discussed later in this article.

Recent direct calorimetric studies (Carpenter *et al.* 1986) indicate that incommensurate behaviour in mineral solid solutions is generally associated with large stabilization enthalpies. Calorimetric data for the plagioclase structures indicate stabilization enthalpies of the order of kilocalories per mole and in the case of mullite the stabilization enthalpy may be estimated at approximately 10 kilocalories per mole. Clearly incommensurate behaviour is of pivotal importance in rendering these phases stable with respect to disordered solid solutions. We note that these stabilization enthalpies are vastly in excess of those observed in simple compounds such as quartz and sodium nitrite and reflect the role of chemical variables on the ordering behaviour of these materials.

In the study of both crystal structure and stabilization enthalpies in incommensurate mineral solid solutions computer simulation is now a viable tool and such studies are likely to become even more important in the immediate future.

3. Incommensurate structure in compounds

(*a*) *Quartz*

The incommensurate behaviour of quartz has as long been the subject of detailed study (Aslanyan *et al.* 1983; Dolino *et al.* 1989). This occurs in a very narrow temperature interval at the β to α transition temperature at 573 °C (846 K). The incommensurate structure is associated with the existence of an improper Lifshitz invariant involving a soft optical and an acoustical mode. High quartz (the β form) is hexagonal with space group $P6_222$, and the low temperature commensurate structure (the α phase) is trigonal with space group $P3_221$. Both the β–α, and the incommensurate phase transition in quartz depend on the dynamic properties of the SiO_4 tetrahedra and their linkages which produce a fully three-dimensional framework structure. From data on the optical frequencies of quartz we know that the weakest forces in the structure relate to Si–O–Si bending. These tend to leave the individual SiO_4 tetrahedra undistorted (giving rise to the so-called rigid modes). Deformations of the individual tetrahedra involving O–Si–O bending are associated with somewhat higher frequencies, but with force constants some 6% of those associated with the Si–O stretching modes which define the highest optical frequencies. These bonding criteria provide the general constraints on transformation mechanisms in quartz, and serve, when taken together with a symmetry analysis of the problem, to establish the nature of the incommensurate behaviour.

First we consider the full symmetry of the components in the modulation and start by studying a single SiO_4 tetrahedron. In the β structure above 573 °C each tetrahedron is located on a site with site symmetry 222. The three diad axes may be chosen in directions x (parallel to the modulation wave vector and a^*), y, and with z parallel to the screw hexad axis. In the transformation to the α structure, which we must regard as the dominant structure in the incommensurate phase, a quasi-rigid-body rotation about the y axis occurs. Above the β–α transition this gives rise to a soft mode along ζ00. Thus in the transition the diad parallel to the y axis is retained but both the remaining diad axes parallel to x and z are lost.

We now apply the selection rules to establish the full symmetry of the additional components. Since the symmetry at Q must be retained the diad parallel to x must be odd. Opposite character for symmetry elements that take Q into $-Q$ implies that the diad parallel to y must be odd, and that the diad parallel to z must be even. For $Q = 0$ (the Γ point) this symmetry accords with full space group symmetry $P6_2$. In principle, therefore we are permitted to include on this basis translations parallel to z, rotations of the tetrahedra about z and an xy distortion of the tetrahedra (presumably with O–Si–O bending and minimal Si–O stretching). In the model of the incommensurate structure due to Valade (Heine 1988) the rotation about z is singled out and originates as the gradient of the transverse acoustical mode. Note that the related shear combines with the optical mode associated with the β to α rigid-body rotation to give an improper Lifshitz invariant. It is likely that the other contributions, from z translation and xy distortion, are also important in the relaxation processes which generate the incommensurate structure. The two-component structure theory, as outlined above, permits one to establish the nature of all the required eigenvectors that contribute to the incommensurate structure. Current computer simulation studies have provided important data on these eigenvectors (F. S. Tautz, personal communication 1990).

(b) *Nepheline*

Nepheline is a complex silicate mineral which shows an interesting incommensurate phase transformation at relatively low temperatures (180 °C). Its incommensurate behaviour has not previously been analysed in detail in spite of the fact that much is already known about its structure. It is of theoretical interest since the incommensurate structure involves interaction between two ordering processes both of which require the loss of threefold axes of symmetry and are thus necessarily described initially in terms of irreducible two-dimensional representations. In the incommensurate coupling interaction both degeneracies are split and a second-order phase transition results. This interaction constitutes a new type of improper Lifshitz invariant.

Nepheline is hexagonal with space group $P6_3$ and has been described as a stuffed derivative of the tridymite structure (SiO_2) with Al replacing Si in tetrahedral sites, and Na and K ions added to balance the electrostatic charge associated with the Al for Si substitution. The formula may be written as $K_x[\ ]_{1-x}Na_3\ Al_{4-x}Si_{4+x}O_{16}$, where x is usually about $\frac{1}{3}$ and the symbol [ ] corresponds to a vacancy on a K site. There are two formula units in the small, high-temperature unit cell. The earliest structural studies on natural nepheline by Hahn & Buerger (1955) showed that the oxygen atom O_1, which should lie on the triad axis in space group $P6_3$, is permanently displaced, and appears in X-ray structural analyses with weight one-third in triad related positions just off the threefold axis. Subsequent structural studies on a number of nephelines by Dollase (1970) and Parker (1972) and studies made at high temperature by Foreman & Peacor (1970) confirmed this interesting feature of the

Table 1. *Representations for the reciprocal vector* $\frac{1}{3}, \frac{1}{3}, 0$ *in space groups* $P6_3$, $\omega = \exp(2\pi i/3)$

	$\{E/0\}$	$\{E/t\}$	$\{E/t^2\}$	$\{C_{3^+}/0\}$	$\{C_{3^+}/t\}$	$\{C_{3^+}/t^2\}$	$\{C_{3^-}/0\}$	$\{C_{3^-}/t\}$	$C_{3^-}/t^2\}$
E_1	1	ω	ω^*	1	ω	ω^*	1	ω	ω^*
	1	ω^*	ω	1	ω^*	ω	1	ω^*	ω
E_2	1	ω	ω^*	ω	ω^*	1	ω^*	1	ω
	1	ω^*	ω	ω^*	ω	1	ω	1	ω^*
E_3	1	ω	ω^*	ω^*	1	ω	ω	ω^*	1
	1	ω^*	ω	ω	1	ω^*	ω^*	ω	1
η	2	-1	-1	2	-1	-1	2	-1	-1
ξ	0	-1	1	0	-1	1	0	-1	1

structure. Natural nepheline has a resulting incommensurate structure at low temperature associated with the ordering of the displaced O_1 atoms and shows satellite intensity maxima paired about the points $\pm(\frac{1}{3}, \frac{1}{3}, 0)$ in the hexagonal reciprocal cell at fractional coordinates of approximately ± 0.20 along c^* (McConnell 1962). These O_1 displacement vectors for the oxygen atom O_1 define a basis for a three-dimensional three-state Potts model that is rather closely related to the classical example of the adsorption of He on Grafoil since it requires a threefold increase in the volume of the unit cell on ordering.

It is possible to define the corresponding supercell and space group for O_1 ordering uniquely from the following argument. The vector $\frac{1}{3}, \frac{1}{3}, 0$ has threefold symmetry only and we are required to augment the point group symmetry of this vector with translations t and t^2 associated with the tripling of the unit cell volume. The relevant irreducible representations of the vector $\frac{1}{3}, \frac{1}{3}, 0$ comprise three two-dimensional representations which conform individually to retention of each of the three triad axes in the small $P6_3$ cell in turn as indicated in table 1. The observed pattern of displacements of O_1 imply that two of the two-dimensional irreducible representations (E_2 and E_3) are irrelevant and we are forced to conclude that the supercell must retain the triad axis at the cell origin. Hence it is inevitable that we describe the simple transformation associated with O_1 ordering in terms of the two-dimensional representation E_1. In describing this ordering it is necessary to write down two real representations which are derived from the appropriate complex conjugate representations as indicated in table 1. These real functions may be used to prepare difference functions which must be used in describing the ordered structure (McConnell & Heine 1984, 1985*a*). Note that the choice of the representation with order parameter η corresponds to normal ordering where complete order is achieved within a supercell with symmetry $P6_3$, and that the second representation ξ relates to a state of partial order only.

So far we have established one of the two possible component structures in the incommensurate structure. The nature of the second component structure was established (McConnell 1981*b*) from heating experiments in which the time–temperature dependence of the loss of incommensurate satellite intensity was monitored. These experiments showed that the disordering of the incommensurate structure had an associated activation energy of 0.84 eV. This was identified with the disordering of vacancies through the migration of potassium ions along the prominent channels in the nepheline structure, thus endorsing previous suggestions that potassium-vacancy ordering had some bearing on the development of the incommensurate structure (Foreman & Peacor 1970). Having established that

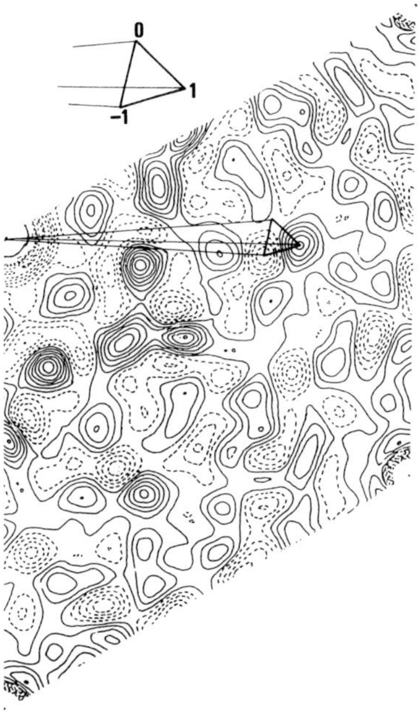

Figure 1. Difference Patterson projection on 0001 within the small $P6_3$ cell showing the vectors, with weights 1, −1 and 0, between the potassium-vacancy modulation and the O_1 displacements about the triad axis.

potassium-vacancy ordering as the second component structure in the modulation it remained to establish its full symmetry. Since there is only one acceptable irreducible representation for the vector $\frac{1}{3}, \frac{1}{3}, 0$ it is obvious that we must again use the two-dimensional representation E_1 of table 1 to describe the potassium-vacancy ordering.

Finally we establish the condition for interaction, and the phase relationships between the two ordering events. First we note that it would be possible to combine these two ordering events in phase to generate a simple $P6_3$ supercell with tripled unit cell volume. However, this is clearly not the correct solution. Since the interaction gives rise to an incommensurate structure we may use the gradient invariant concept to establish the necessary selection rules. There are two possibilities, and both necessarily involve the loss of degeneracy. In the first solution the real representation η can be used for the O_1 displacements, and the second real representation ξ, associated with partial order, can be used for the potassium-vacancy component. In the second acceptable solution this assignment is reversed. In either case we envisage that the two components of the modulation must exist in quadrature along the wave vector parallel to the c axis. Data from an experimentally derived difference Patterson function were used to establish the validity of this argument and these have been illustrated in figure 1. In this projection vectors labelled 1, −1 and 0 correspond to the products of difference functions for O_1 displacements and potassium-vacancy ordering and these characters, as described in McConnell & Heine (1985 *a*), are compatible with either of the two solutions deduced above.

This analysis of the nature of the incommensurate structure in nepheline is particularly interesting in that it represents a new type of interaction, illustrates the power of the methods available in component structure theory, and finally, that it illustrates the fact that an incommensurate interaction can be responsible for splitting degeneracies in a Potts model system to yield a second-order transition where the normal possibility relates to first-order transformation behaviour.

4. Incommensurate structure in solid solutions

In §2 dealing with the theory of incommensurate structures it was noted that the degree of rigour possible in the use of the two-component structure theory is dependent particularly on whether or not the system involves simple compounds or complex solid solutions. In the latter case it was noted that the ideal component structures may differ both in symmetry, and in ordering capability, as a function of chemical composition. Two such structural components with differing ordering characteristics may exist within an incommensurate modulation and interact favourably if the gradient invariant concept is also satisfied, i.e. if we can establish the existence of an improper Lifshitz invariant. In the case of mullite, which was used as an example of this phenomenon, the existence of space groups of appropriate symmetry, which order Al and Si in tetrahedral sites and vacancies respectively, leads to an extremely stable structure which orders with a very large reduction in enthalpy which provides a substantial stabilization energy.

Not all of the known incommensurate solid solutions have the simplicity of the mullite example and it is now apparent that real incommensurate solid solutions range from those like mullite to examples where the concept of component structures is less easy to justify. The incommensurate structure of the plagioclase feldspars is of the latter type and we have chosen to discuss it here for two reasons. Firstly the incommensurate behaviour of the plagioclases has been known since the 1950s (Bown & Gay 1958) and there have been numerous attempts, both theoretical and practical, to establish the structure of the incommensurate phase, and the *raison d'etre* of the phenomena. The feldspars are also extremely common rock forming minerals and make up a very large fraction of the Earth's crust.

The plagioclase feldspars comprise triclinic solid solutions with the general formula $Na_xCa_{1-x}Al_{2-x}Si_{2+x}O_8$ with end member structures $NaAlSi_3O_8$ (albite) and $CaAl_2Si_2O_8$ (anorthite). They are framework structures like quartz and nepheline, with Si and Al in tetrahedrally coordinated sites. The ordering of Al and Si dominates the transformation behaviour of the feldspars. It is convenient to illustrate the structure and Al/Si ordering of both feldspar end members by drawing out a section parallel to the $\bar{2}01$ plane, as shown in figure 2, which shows the fourfold rings of tetrahedra that are characteristic of all feldspar structures. The planes of fourfold rings are stacked one above the other with counter rotation of the fourfold rings in such a way that a fully dimensionally linked framework structure results. Since the feldspars are triclinic in structure the incommensurate wave vector is not restricted in orientation and it is known to change its orientation and wavelength continuously as a function of composition between the sodium and calcium ends of this binary system (Bown & Gay 1958). Near the sodium end of the system the wave vector is approximately parallel to the *c* axis, it has a short wavelength and the intensity maxima are always diffuse. At the calcium end of the system the satellite intensity maxima are paired about the position of a simple superlattice, the reduced wave vector Q is approximately parallel to the *b* crystallographic axis and the maxima are extremely sharp defining a wavelength of the order of 60 Å (6 nm). The incommensurate structure itself exsolves into two incommensurate structures in a narrow com-

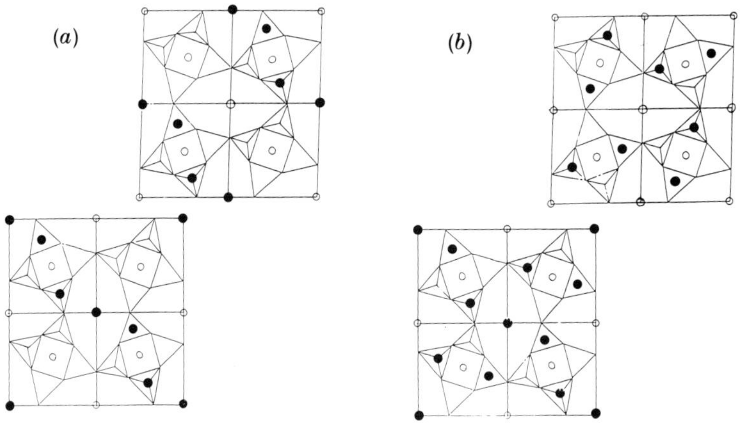

Figure 2. Diagrams illustrating the Si/Al ordering schemes in the end-member structures for (*a*) albite and (*b*) anorthite. In each case the tetrahedra containing Al are shown with black dots. Each diagram shows two adjacent $\bar{2}01$ layers of the structure. In the albite structure such pairs of layers are identical and lattice points occur in both. In the anorthite sequence the ordering pattern renders successive layers inequivalent and this leads to a doubling of the unit cell volume for anorthite.

positional interval in the middle of this binary system and the equilibrium microstructures in this compositional range have a fixed period and produce a very distinctive iridescence or schiller that was first studied seriously by Lord Rayleigh in 1923.

We now attempt an analysis of the origin of incommensurate structure in these materials. We first establish the ordering and transformation behaviour for the system. Here the most important factor to note is that the end member compounds albite and anorthite both have ordered Si/Al distributions but that these are not isostructural. The details are illustrated in figure 2. Since the Al/Si ratio in anorthite is one to one there is effectively only one possible scheme for ordering which requires that each Al containing tetrahedron in the framework is surrounded by four neighbouring tetrahedra containing Si atoms. This ordering scheme results in a doubling of the unit cell volume in anorthite and takes place at temperatures at or near to the melting point at 1500 °C.

Although this scheme is acceptable for ordering in the sodium end member albite with a ratio of Al to Si of one to three, it is not the ordering pattern observed. Ordering in albite takes place in such a way that the cell volume remains the same as for the disordered solid solution, and occurs at a temperature of *ca.* 675 °C. In discussing the ordering schemes for anorthite and albite there is one additional factor which turns out to be of crucial importance. In the energetics of ordering Si and Al in the feldspars it is a long-established principle that the primary contribution to disorder enthalpy derives from the existence of Al neighbours in adjacent tetrahedra. It is also very easy to show that any addition of calcium feldspar to the ordered albite structure (with substitution of Al for Si) inevitably violates this principle. These several facts serve to establish the origin of the incommensurate structure in the plagioclase feldspars since we must consider the ordering behaviour within the solid solution in terms of the two radically different ordering schemes associated with the end-member structures. For compositions close to albite in composition the extra Al atoms are incorporated by creating a local ordering defect that is compatible with the anorthite ordering scheme. This requires that Al be displaced from its normal site in the ordered albite structure, and structural data provide concrete evidence for this effect (Ribbe 1983). Incommensurate wavelength depends on the fact that the defects have a tendency to cluster in planes in an effort to minimize the overall strain energy. A similar analysis may be used to elucidate the nature of the incommensurate structures in binary solid solution members rich in anorthite. Here the predominant structural influence relates to the development of the anorthite ordering scheme. The gradient ploy may be used to demonstrate that Na and Ca ions, and the extra Si atoms, are ordered in quadrature with the primary anorthite ordering scheme (McConnell 1978). Calorimetric data from Carpenter *et al.* (1985) demonstrate that, in this compositional range, and at 65% anorthite, the enthalpy of the incommensurate structure is close to that of a mechanical mixture of the ordered plagioclase end members implying that an extremely high degree of order exists in the incommensurate structure.

This brief review of incommensurate behaviour in solid solutions confirms that this is an extremely important class of behaviour leading to very substantial ordering, and related enthalpy stabilization in solid solutions.

References

Aslanyan, T. A., Levanyuk, A. P., Vallade, M. & Lajzerowicz, J. 1983 Various possibilities for formation of incommensurate superstructure near the α–β transition in quartz. *J. Phys.* C**16**, 6705–6712.

Axe, J. D., Harada, I. & Shirane, G. 1970 Anomalous acoustic dispersion in centrosymmetric crystals with soft optic phonons. *Phys. Rev.* B**1**, 1227–1234.

Bandour, J. L. & Sanquer, M. 1983 Structural phase transitions in polyphenyls. *Acta crystallogr.* B**39**, 75–84.

Bown, M. G. & Gay, P. 1958 The reciprocal lattice geometry of the plagioclase feldspar structures. *Z. Kristallogr. Kristallgeom.* **111**, 1–14.

Cameron, W. E. 1977 Mullite: a substituted alumina. *Am. Min.* **62**, 747–755.

Carpenter, M. A., McConnell, J. D. C. & Navrotsky, A. 1985 Enthalpies of ordering in the plagioclase feldspar solid solution. *Geochim. cosmochim. Acta* **49**, 947–966.

Dolino, G., Berge, B., Vallade, M. & Moussa, F. 1989 Inelastic scattering studies of the origin of the incommensurate phase of quartz. *Physica* B**156**, 15–16.

Dollase, W. A. 1970 Least squares refinement of the structure of a plutonic nepheline. *Z. Kristallogr. Kristallogeom.* **132**, 27–44.

Foreman, N. & Peacor, D. R. 1970 Refinement of the nepheline structure at several temperatures. *Z. Kristallogr. Kristallogeom.* **132**, 45–70.

Gartstein, E., Mason, T. O. & Cohen, J. B. 1986 Defect agglomeration in wustite at high temperatures. I. The defect arrangement. *J. Phys. Chem. Solids* **47**, 759–773.

Hahn, T. & Buerger, M. J. 1955 The detailed structure of nepheline $KNa_3Al_4Si_4O_{16}$. *Z. Kristallogr. Kristallogeom.* **106**, 308–338.

Heine, V. 1988 In *Physical properties and thermodynamic behaviour of minerals* (ed. E. K. H. Salje) pp. 1–15. Dordrecht: Reidel.

Heine, V. & McConnell, J. D. C. 1981 Original of incommensurate phases in insulators. *Phys. Rev. Lett.* **46**, 1092–1095.

Heine, V. & McConnell, J. D. C. 1984 The origin of incommensurate structures in insulators. *J. Phys.* C**17**, 1199–1220.

Landau, L. D. & Lifshitz, E. M. 1958 *Statistical physics.* Reading, Massachusetts: Addison Wesley.

Levanyuk, A. P. & Sannikov, D. G. 1976 Phase transitions into inhomogeneous states. *Ferroelectrics* **14**, 643–645.

Lifshitz, E. M. 1942 On the theory of phase transitions of second order. *Soviet J. Phys.* **6**, 61–73; 251–263.

McConnell, J. D. C. 1962 Electron diffraction study of subsidiary maxima of scattered intensity in nepheline. *Mineralog. Mag.* **33**, 114–124.

McConnell, J. D. C 1978 The intermediate plagioclase feldspars: an example of a structural resonance. *Z. Kristallogr. Kristallgeom.* **147**, 45–62.

McConnell, J. D. C. 1981*a* Electron-optical study of modulated mineral solid solutions. *Bull. Mineral.* **104**, 231–235.

McConnell, J. D. C. 1981*b* Time-temperature study of the intensity of satellite reflections in nepheline. *Am. Miner.* **66**, 990–996.

McConnell, J. D. C. & Heine, V. 1984 An aid to the structural analysis of incommensurate phases. *Acta crystallogr.* A**40**, 473–482.

McConnell, J. D. C. & Heine, V. 1985*a* The symmetry properties of difference Patterson functions. *Acta crystallogr.* A**41**, 382–386.

McConnell, J. D. C. & Heine, V. 1985*b* Incommensurate structure and stability of mullite. *Phys. Rev.* B**31**, 6140–6142.

Parker, J. M. 1972 The domain structure of nepheline. *Z. Kristallogr. Kristallgeom.* **136**, 255–272.

Rayleigh, Lord 1923 Studies of iridescent colour and the structures producing it. *Proc. R. Soc. Lond.* A**103**, 34–45.

Ribbe, P. H. 1983 In *Feldspar mineralogy*, pp. 21–55. *Reviews in Mineralogy* vol. 2. Min. Soc. Amer.

Sannikov, D. G. & Golovko, V. A. 1989 Role of gradient invariants in the theory of the incommensurate phase. *Sov. Phys. solid St.* **31**, 137–140.

Discussion

P. W. Anderson (*Princeton, U.S.A.*) Is there not a close relation to the 'blue phases' of cholesteric liquid crystals? Here there are two possible directions of twist and they combine to make a lattice of defects, over a small temperature range near the cholesteric–isotropic transition.

J. D. C. McConnell. This seems quite likely but might be difficult to prove.

J. N. Murrell (*University of Sussex, U.K.*). Are there any aspects of the Jahn–Teller effect which enter the formation of incommensurate structures?

J. D. C. McConnell. Yes, we have studied an incommensurate complex copper nitrate phase in which the Jahn–Teller effect forms the basis of its incommensurate behaviour.

J. M. Lynden-Bell (*Cambridge University Chemical Laboratories, U.K.*). In the last example the relevant irreducible representation is two dimensional. Couldn't there have been a proper Lifschitz invariant and, if so, why wasn't this important?

J. D. McConnell. The representation E_1 has a third-order invariant and thus leads to first-order transformation which splits the degeneracy. The Lifshitz invariant is not relevant.

A. O. E. Animalu (*University of Nigeria, Nsukka, Nigeria*). Is the transition from commensurate to incommensurate structure always continuous? Professor McConnell's answer for getting 'order' out of 'chaos' in natural selection implies that the emergence of a new species that is incommensurate with the old species may be spontaneous (i.e. discontinuous) rather than smooth (i.e. continuous) contrary to what second-order phase transition would predict.

J. D. C. McConnell. All known examples appear to be second order.

B. Coles (*Imperial College, London, U.K.*). A category of incommensurate structures perhaps closer to Professor McConnell's minerals are the long-period stacking fault structures in certain metallic alloys. Even metallic samarium can be treated as nine-layer close-packed structure produced by stacking faults every third layer in the simpler close-packed structures.

J. D. C. McConnell. Where the period is incommensurate this is likely to be true, as in CuAu.

Bonding and structure of intermetallics: a new bond order potential

By D. G. Pettifor and M. Aoki

Department of Mathematics, Imperial College of Science, Technology and Medicine, London SW7 2BZ, U.K.

Intermetallics such as the transition metal aluminides present theorists with a challenge since bonding is not well described by currently available pair or embedded atom potentials. We show that a new angularly dependent, many-body potential for the bond order has all the necessary ingredients for an adequate description. In particular, by linearizing the moment-recursion coefficient relations, a cluster expansion is derived which is applicable to any lattice and chemical ordering and which allows a derivation of the earlier ring ansatz. It can account for both the negative Cauchy pressure of cubic metals and the oscillatory behaviour across the transition metal aluminide series of the three-body cluster interaction Φ_3.

1. Introduction

Intermetallics have come to the fore during the past few years with the realization that polycrystalline Ni_3Al could be ductilized by adding very small amounts of boron (Aoki & Izumi 1979). The search for new intermetallics for use at high temperatures, in jet engines for example, has focused on the transition metal aluminides which are often both light and oxidation resistant (Dimiduk & Miracle 1989). In particular, alloy designers have been interested in whether it is possible to ductilize the tri-aluminides Al_3T by alloying so that their crystal structure changed from tetragonal (and brittle) to cubic (and hopefully ductile).

Structure maps, which order the known structural data base on binary compounds within a limited number of two-dimensional or three-dimensional plots, can provide a useful guide in the search for new pseudo-binary alloys with a required structure type (Pettifor 1991). Figure 1 shows the relevant part of the AB_3 structure map which has ordered 'the wood from the trees' by characterizing each element in the periodic table with a single phenomenological coordinate, called the Mendeleev number $\mathscr{M}$ (Pettifor 1988). We see that the intermetallics Al_3Hf, Al_3Ti, Al_3Ta, Al_3Nb, and Al_3V with Mendeleev numbers ranging from $\mathscr{M}_A$ equals 50 to 53 respectively all fall within the tetragonal DO_{22} domain. It was known that Al_3Ti could be stabilized in the cubic $L1_2$ crystal structure by replacing some of the aluminium with Cu, Ni, or Fe. The structure map suggests that it might be possible to stabilize the other tetragonal tri-aluminides in the cubic crystal structure by alloying so that the average Mendeleev number for the B sites moves down into the cubic $L1_2$ domain.

This hope has only been partially realized. Schneibel & Porter (1989) have indeed succeeded in stabilizing cubic Al_3Zr by alloying to take the average Mendeleev number $\bar{\mathscr{M}}_B$ down into the cubic domain. However, attempts to stabilize cubic Al_3Nb

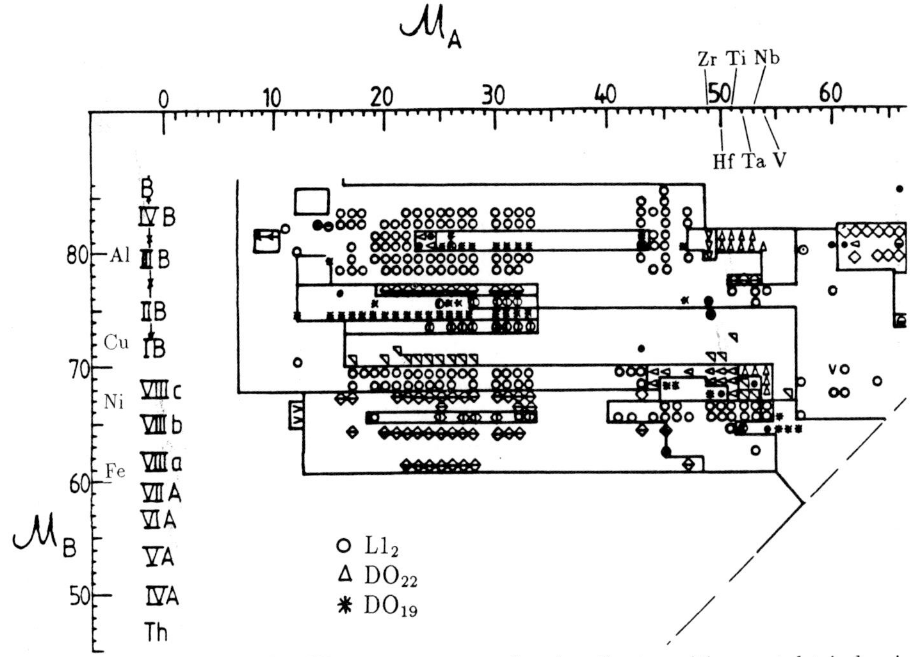

Figure 1. Relevant part of the AB_3 structure map showing the transition metal tri-aluminides. (After Nicholson *et al.* 1989.)

have failed (Subramanian *et al.* 1989). Moreover, the cubic tri-aluminides based on Al_3Zr and Al_3Ti still remain brittle, cleaving transgranularly, even though they have the same crystal structure as ductile single crystals of Cu_3Au or Ni_3Al. Theorists, are, therefore, faced with two immediate problems regarding the transition metal tri-aluminides: (i) Why if the cubic close-packed phase can be stabilized, does it remain brittle? (ii) Why can some tetragonal tri-aluminides be stabilized in the cubic form (e.g. Al_3Ti) but others cannot (e.g. Al_3Nb)? We shall see that the answer to both questions requires a proper quantum mechanical treatment of the bonding at the atomistic level.

2. A new bond order potential

A clue to the origin of the brittleness of the cubic transition metal tri-aluminides is provided by their elastic constants. Fu (1991) has recently calculated within first principles local density functional theory that cubic Al_3Ti has a Cauchy pressure $C_{12}-C_{44}$ of -0.08×10^{11} N m^{-2}, which is to be compared with a Cauchy pressure of $+0.13\times10^{11}$ N m^{-2} for Ni_3Al. This has important consequences for the nature of the bonding at the atomistic level. If the bonding is describable by nearest-neighbour pairwise potentials such as Lennard-Jones, then the Cauchy pressure would be zero. If the bonding is more metallic in that spherical atoms are embedded in the electron gas of the surrounding neighbours, then the Cauchy pressure would be positive (Johnson 1988). A negative Cauchy pressure puts cubic Al_3Ti in the same class as the four-fold coordinated semi-conductor Si with a $C_{12}-C_{44}$ of -0.16×10^{11} N m^{-2}. This implies that angularly dependent many-body forces are playing a crucial role in the transition metal tri-aluminides.

Recently a new many-body potential for the bond order has been proposed which explicitly includes the angular character of the bonding orbitals (Pettifor 1989,

1990). It is derived from tight binding Hückel theory in which the quantum mechanical bond energy between a given pair of atoms i and j is written in the chemically intuitive form as follows:

$$U^{ij}_{\text{bond}} = 2h(R_{ij})\,\Theta_{ij}, \tag{2.1}$$

where $h(R_{ij})$ is the appropriate σ, π or δ bond integral between atoms i and j a distance R_{ij} apart (see eqs (65) and (80) of Pettifor (1990)) and Θ_{ij} is the corresponding bond order which is defined as the difference between the number of electrons of a given spin in the bonding $\frac{1}{\sqrt{2}}|\phi_i+\phi_j\rangle$ and anti-bonding $\frac{1}{\sqrt{2}}|\phi_i-\phi_j\rangle$ states. The factor 2 in equation (2.1) accounts for spin degeneracy. The bonding between any given pair of atoms will, of course, be weakened by the presence of bonds with other neighbouring atoms. Thus the bond order is not pairwise but is dependent on the surrounding atomic environment.

Its analytic dependence may be obtained by using the recursion method of Haydock *et al.* (1972) to write the bond order as an integral over the difference of two continued fractions:

$$\Theta_{ij} = -\frac{1}{\pi}\,\text{Im}\int^{E_{\text{F}}} [G^{+}_{00}(E) - G^{-}_{00}(E)]\,\text{d}E, \tag{2.2}$$

where E_{F} is the Fermi energy and Im is the imaginary part of the bonding and antibonding Green's functions which are given by

$$G^{\pm}_{00}(E) = \langle u^{\pm}_0|(E-H)^{-1}|\,u^{\pm}_0\rangle$$

$$= \frac{1}{(E-a^{\pm}_0)-(b^{\pm}_1)^2/(E-a^{\pm}_1)-\ldots}, \tag{2.3}$$

where $|u^{\pm}_0\rangle = \frac{1}{\sqrt{2}}|\phi_i \pm \phi_j\rangle$. The coefficients are determined by the Lanczos recursion algorithm, namely

$$b^{\pm}_{n+1}\,|u^{\pm}_{n+1}\rangle = H\,|u^{\pm}_n\rangle - a^{\pm}_n\,|u^{\pm}_n\rangle - b^{\pm}_n\,|u^{\pm}_{n-1}\rangle \tag{2.4}$$

with the boundary condition that $|u^{\pm}_{-1}\rangle$ vanishes. The hamiltonian H is, therefore, tridiagonal with respect to the recursion basis $|u^{\pm}_n\rangle$, having non-zero elements

$$\langle u^{\pm}_n|\,H\,|u^{\pm}_n\rangle = a^{\pm}_n \tag{2.5}$$

and

$$\langle u^{\pm}_{n+1}|\,H\,|u^{\pm}_n\rangle = b^{\pm}_{n+1}. \tag{2.6}$$

Thus the hamiltonian with respect to the bonding and antibonding recursion basis may be characterized by the semi-infinite linear chain with site diagonal elements $a^{\pm}_n$ and intersite hopping matrix elements $b^{\pm}_{n+1}$, namely

$$\begin{array}{ccccccccccc} a^{\pm}_0 & & a^{\pm}_1 & & a^{\pm}_2 & & & a^{\pm}_{n-1} & & a^{\pm}_n & \\ \times & -b^{\pm}_1- & \times & -b^{\pm}_2- & \times & -\cdots- & & \times & -\;b^{\pm}_n- & \times & -\cdots. \\ 0 & & 1 & & 2 & & & n-1 & & n & \end{array}$$

A many-body form for the bond order may be derived by performing perturbation theory with respect to the average semi-infinite linear chain, namely

$$\begin{array}{ccccccccccc} \bar{a}_0 & & \bar{a}_1 & & \bar{a}_2 & & & \bar{a}_{n-1} & & \bar{a}_n & \\ \times & -\bar{b}_1- & \times & -\bar{b}_2- & \times & -\cdots- & & \times & -\bar{b}_n- & \times & -\cdots, \\ 0 & & 1 & & 2 & & & n-1 & & n & \end{array}$$

where $\bar{a}_n = \frac{1}{2}(a_n^+ + a_n^-)$ and $\bar{b}_n = \frac{1}{2}(b_n^+ + b_n^-)$. It follows from the first-order Dyson equation that

$$G_{00}^{\pm} = G_{00}^0 \pm \sum_{n=0}^{\infty} G_{0n}^0 G_{n0}^0 \,\delta a_n \pm 2 \sum_{n=1}^{\infty} G_{0(n-1)}^0 G_{n0}^0 \,\delta b_n, \tag{2.7}$$

where $\delta a_n = \frac{1}{2}(a_n^+ - a_n^-)$, $\delta b_n = \frac{1}{2}(b_n^+ - b_n^-)$, and G^0 is the Green's function for the average semi-infinite linear chain. Substituting into (2.2) the bond order becomes

$$\Theta = -2\left[\sum_{n=0}^{\infty} \chi_{0n,\,n0}(E_F)\,\delta a_n + 2\sum_{n=1}^{\infty} \chi_{0(n-1),\,n0}(E_F)\,\delta b_n\right], \tag{2.8}$$

where the response functions $\chi_{0m,\,n0}(E_F)$ are defined by

$$\chi_{0m,\,n0}(E_F) = \frac{1}{\pi}\,\mathrm{Im}\int^{E_F} G_{0m}^0(E)\,G_{n0}^0(E)\,\mathrm{d}E. \tag{2.9}$$

The coefficients δa_n, δb_n may be written in terms of the local topology about the bond by using the well-known relationship between the recursion coefficients $a_n^{\pm}$, $b_n^{\pm}$ and the moments $\mu_n^{\pm} = \langle u_0^{\pm}|H^n|u_0^{\pm}\rangle$, namely

$$\mu_0^{\pm} = 1, \tag{2.10}$$

$$\mu_1^{\pm} = a_0^{\pm}, \tag{2.11}$$

$$\mu_2^{\pm} = (a_0^{\pm})^2 + (b_1^{\pm})^2, \tag{2.12}$$

$$\mu_3^{\pm} = (a_0^{\pm})^3 + 2a_0^{\pm}(b_1^{\pm})^2 + a_1^{\pm}(b_1^{\pm})^2, \tag{2.13}$$

$$\mu_4^{\pm} = (a_0^{\pm})^4 + 3(a_0^{\pm})^2(b_1^{\pm})^2 + 2a_0^{\pm}a_1^{\pm}(b_1^{\pm})^2 + (a_1^{\pm})^2(b_1^{\pm})^2 + (b_1^{\pm})^2(b_2^{\pm})^2 + (b_1^{\pm})^4 \tag{2.14}$$

and

$$\begin{aligned}\mu_5^{\pm} = {} & (a_0^{\pm})^5 + 4(a_0^{\pm})^3(b_1^{\pm})^2 + 3(a_0^{\pm})^2 a_1^{\pm}(b_1^{\pm})^2 \\ & + 3a_0^{\pm}(b_1^{\pm})^4 + 2a_0^{\pm}(a_1^{\pm})^2(b_1^{\pm})^2 + 2a_0^{\pm}(b_1^{\pm})^2(b_2^{\pm})^2 \\ & + 2a_1^{\pm}(b_1^{\pm})^4 + 2a_1^{\pm}(b_1^{\pm})^2(b_2^{\pm})^2 + (a_1^{\pm})^3(b_1^{\pm})^2 \\ & + a_2^{\pm}(b_1^{\pm})^2(b_2^{\pm})^2.\end{aligned} \tag{2.15}$$

The difference between the bonding and antibonding moments may be displayed explicitly by writing

$$\mu_n^{\pm} = \mu_n \pm \zeta_{n+1}, \tag{2.16}$$

where it follows that since $|u_0^{\pm}\rangle = \frac{1}{\sqrt{2}}|\phi_i + \phi_j\rangle$ we have that μ_n is the average nth moment with respect to the appropriate orbitals on site i and j, namely

$$\mu_n = \tfrac{1}{2}[\langle\phi_i|H^n|\phi_i\rangle + \langle\phi_j|H^n|\phi_j\rangle] = \tfrac{1}{2}(\mu_n^i + \mu_n^j) \tag{2.17}$$

and ζ_{n+1} is the interference term, namely

$$\zeta_{n+1} = \langle\phi_i|H^n|\phi_j\rangle. \tag{2.18}$$

The interference terms control the bond order which can be seen using the moment expansion for $G^{\pm}$, namely

$$G^+(Z) - G^-(Z) = \sum_{n=0}^{\infty} \frac{\mu_n^+ - \mu_n^-}{Z^{n+1}} = 2\sum_{n=1}^{\infty} \frac{\zeta_{n+1}}{Z^{n+1}}. \tag{2.19}$$

However, the moment expansion is notoriously ill conditioned. We, therefore, rewrite the stable recursion expansion for the bond order, equation (2.8), in terms of the ζ_{n+1} by linearizing the moment-recursion coefficient relations in equations (2.10)–(2.15). We find to first order in $\zeta_{n+1}/\mu_2^{n/2}$, taking $\bar{a}_0 = \mu_1 = 0$ as the reference energy, that the average recursion coefficients $\bar{a}_n, \bar{b}_n$ are given by

$$\bar{a}_0 = \mu_1 \equiv 0, \tag{2.20}$$

$$\bar{b}_1 = \mu_2^{\frac{1}{2}}, \tag{2.21}$$

$$\bar{a}_1 = \mu_3/\mu_2, \tag{2.22}$$

$$\bar{b}_2 = (\mu_4/\mu_2 - \mu_3^2/\mu_2^2 - \mu_2)^{\frac{1}{2}}, \tag{2.23}$$

and

$$\bar{a}_2 = [\mu_5 - 2\mu_3(\mu_4/\mu_2 - \mu_3^2/\mu_2^2) - \mu_3^3/\mu_2^2]/(\bar{b}_1)^2(\bar{b}_2)^2. \tag{2.24}$$

That is, to first order, the average coefficients $\bar{a}_n, \bar{b}_n$, which enter the reference response functions $\chi_{0m,\,n0}(E_{\mathrm{F}})$, are determined solely by the average moments μ_n which characterize the average Green's function associated with the appropriate orbitals on sites i and j, namely

$$\bar{G}(E) = \tfrac{1}{2}[G_{ii}(E) + G_{jj}(E)]. \tag{2.25}$$

The coefficients $\delta a_n, \delta b_n$, on the other hand, are given to first order in ζ_{n+1} by

$$\delta a_0 = \zeta_2 \equiv \langle \phi_i | H | \phi_j \rangle, \tag{2.26}$$

$$\delta b_1 = \zeta_3/(2\mu_2^{\frac{1}{2}}), \tag{2.27}$$

$$\delta a_1 = \zeta_4/\mu_2 - (\mu_3/\mu_2^2)\,\zeta_3 - 2\zeta_2, \tag{2.28}$$

$$\delta b_2 = [\zeta_5 - (2\mu_3/\mu_2)\,\zeta_4 - (\mu_4/\mu_2 - 2\mu_3^2/\mu_2^2 + \mu_2)\,\zeta_3 + 2\mu_3\,\zeta_2]/[2(\bar{b}_1)^2\,\bar{b}_2], \tag{2.29}$$

and

$$\delta a_2 = \Bigg\{\zeta_6 - \frac{\mu_5 - 2\mu_2\mu_3 - \mu_3^3/\mu_2^2}{(\bar{b}_1)^2(\bar{b}_2)^2}\zeta_5 - \left[\frac{2\mu_4}{\mu_2} + \frac{\mu_3^2}{\mu_2^2} + \frac{4\mu_3^2}{\mu_2\bar{b}_2^2} + \frac{2\mu_3^4}{\mu_2^4(\bar{b}_2)^2} - \frac{2\mu_3\mu_5}{\mu_2^2(\bar{b}_2)^2}\right]\zeta_4$$
$$- \left[\frac{\mu_5}{\mu_2} - \frac{\mu_3^3}{\mu_2^3} - \frac{2\mu_3\mu_4}{\mu_2^2} + \left(\frac{\mu_4}{\mu_2} - \frac{2\mu_3^2}{\mu_2^2} + \mu_2\right)\left(\frac{\mu_3^3}{\mu_2^3(\bar{b}_2)^2} + \frac{2\mu_3}{(\bar{b}_2)^2} - \frac{\mu_5}{\mu_2(\bar{b}_2)^2}\right) + 2\mu_3\right]\zeta_3$$
$$- \left[\frac{2\mu_3\mu_5}{\mu_2(\bar{b}_2)^2} + \mu_2^2 - 2\mu_4 - \frac{2\mu_3^2}{\mu_2} - \frac{2\mu_3^4}{\mu_2^3\bar{b}_2^2} - \frac{4\mu_3^2}{(\bar{b}_2)^2}\right]\zeta_2\Bigg\}\Bigg/[(\bar{b}_1)^2(\bar{b}_2)^2]. \tag{2.30}$$

The coefficients $\delta a_n, \delta b_n$ take a particularly transparent form if we assume that $\bar{b}_2 = \bar{b}_1$ and $\mu_{2n+1}/\mu_2^{(2n+1)/2} \ll 1$. It then follows from equations (2.26)–(2.30) that

$$\delta a_0 = \zeta_2, \tag{2.31}$$

$$2\delta b_1 = \zeta_3/\mu_2^{\frac{1}{2}}, \tag{2.32}$$

$$\delta a_1 = (\zeta_4 - 2\mu_2\,\zeta_2)/\mu_2, \tag{2.33}$$

$$2\delta b_2 = (\zeta_5 - 3\mu_2\,\zeta_3)/\mu_2^{\frac{3}{2}}, \tag{2.34}$$

and

$$\delta a_2 = [\zeta_6 - 4\mu_2(\zeta_4 - 2\mu_2\,\zeta_2) - \mu_2^2\,\zeta_2 - 2\mu_4\,\zeta_2]/\mu_2^2. \tag{2.35}$$

The ζ_n can be represented diagrammatically as in figure 2. The first diagram

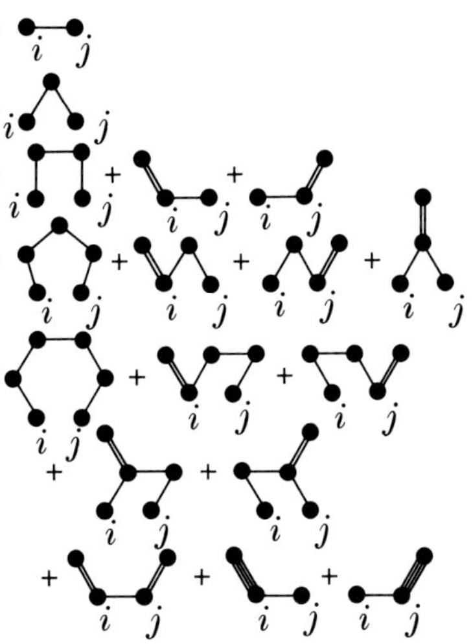

Figure 2. Diagrammatic representation of the interference terms ζ_n between atoms i and j.

represents the ring term ζ_n^r in which all n sites are distinct and there are no self-retracing paths, whereas the latter diagrams are dressed by self-retracing paths which must be summed over all nearest neighbour sites. We have neglected double-counting terms which involve hopping backwards and forwards between atoms i and j. This is a good approximation for s orbitals on lattices with large local coordination z as the hops to the z nearest neighbours swamp the double-counting contribution. It follows from equations (2.31)–(2.35) that the dressed diagrams cancel from the $\delta a_n, \delta b_n$, leaving only the ring terms, namely

$$\delta a_0 = \zeta_2^r, \tag{2.36}$$

$$\delta b_1 = \zeta_3^r / 2\mu_2^{\frac{1}{2}}, \tag{2.37}$$

$$\delta a_1 = \zeta_4^r / \mu_2, \tag{2.38}$$

$$\delta b_2 = \zeta_5^r / 2\mu_2^{\frac{3}{2}}, \tag{2.39}$$

and

$$\delta a_2 = \zeta_6^r / \mu_2^2. \tag{2.40}$$

The response functions consistent with equations (2.36)–(2.40), which were derived by neglecting odd moments and taking $\bar{b}_2 = \bar{b}_1$, are those corresponding to a reference semi-infinite linear chain with $\bar{a}_n = 0$, $\bar{b}_n = \bar{b}_1 \equiv b$. They may be written (Pettifor 1989) as $\chi_{0m,\,n0} = \hat{\chi}_{m+n+2}/|b|$ for $m = n-1$ or n where the reduced susceptibility

$$\hat{\chi}_{m+n+2}(N) = \frac{1}{\pi}\left[\frac{\sin(m+n+1)\,\phi_{\mathrm{F}}}{m+n+1} - \frac{\sin(m+n+3)\,\phi_{\mathrm{F}}}{m+n+3}\right] \tag{2.41}$$

with $\phi_{\mathrm{F}} = \arccos(E_{\mathrm{F}}/2b)$. ϕ_{F} is fixed by the number of valence electrons per spin per bond, N, through

$$N = (2\phi_{\mathrm{F}}/\pi)\,[1 - (\sin 2\phi_{\mathrm{F}})/2\phi_{\mathrm{F}}]. \tag{2.42}$$

Figure 3 shows the behaviour of the first five reduced response functions $\hat{\chi}_n$ as a function of the number of valence electrons per spin per bond. We see that the number of nodes (excluding the end points) equals $(n-2)$.

Figure 3

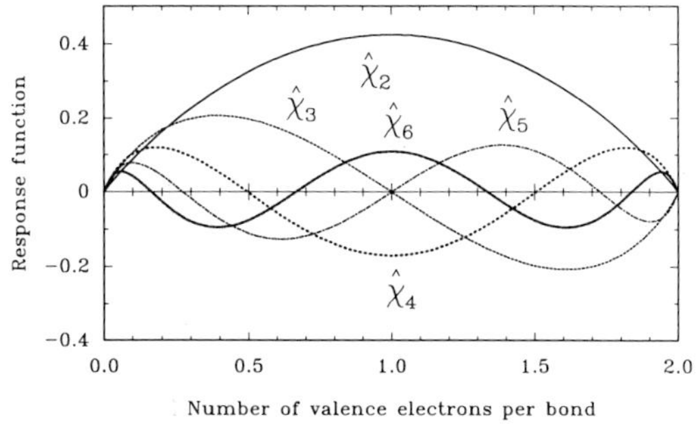

Figure 4

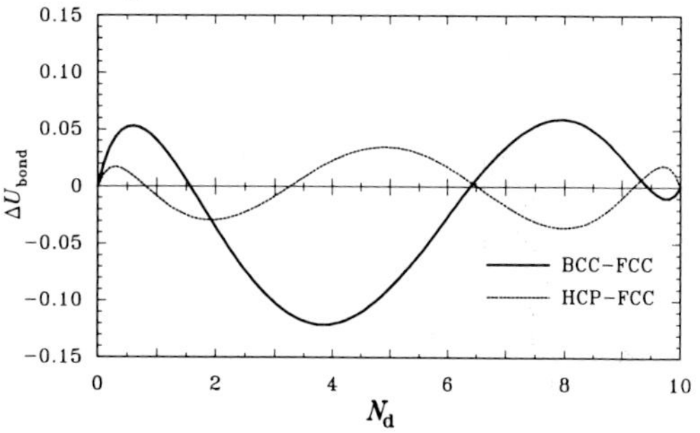

Figure 3. The reduced response functions $\hat{\chi}_n$ as a function of the number of valence electrons per spin per bond N.

Figure 4. The total bond energy difference (in units of band width) between BCC, HCP, and FCC transition metals as a function of the number of valence d electrons per atom N_d for the BCC lattice.

We shall refer to the use of equations (2.36)–(2.41) as the ring approximation (RA), in which the bond order is written

$$\Theta = 2 \sum_{n=2}^{\infty} \hat{\chi}_n(N)\, \zeta_n^r / b^{n-1}. \tag{2.43}$$

Elsewhere we have used the RA potential to investigate the relative stability of linear, square, or tetrahedral s-valent clusters (Pettifor 1989) and the angular character of the embedding function b for sp- and sd-valent systems (Pettifor 1990). Here we demonstrate that it reproduces the observed trends in crystal structure and elastic constants across the transition metal series.

Figure 4 shows the total bond energy difference between the BCC, HCP, and FCC lattices as a function of the number of valence d electrons per atom, retaining the first five terms in (2.43). We seen that it shows the well-known trend from HCP → BCC → HCP → FCC(→ BCC) across the non-magnetic 4d and 5d series (see, for example, fig. 35 of Pettifor (1983)). The BCC stability for nearly half-full bands is associated with the four-membered ring term through $\hat{\chi}_4$, the cubic versus hexagonal stability with the six-membered ring term through $\hat{\chi}_6$.

Figure 5 shows the behaviour of the Cauchy pressure $C_{12} - C_{44}$ as a function of the number of valence d electrons per atom for the BCC lattice. We see that it oscillates in sign, so that the RA potential (unlike pair or embedded atom potentials) can account naturally for the negative Cauchy pressures of brittle metals such as elemental FCC Ir or the cubic pseudo-binary intermetallic Al_3Ti. It remains to fit RA potentials to these specific systems and to model the behaviour of their defects such as dislocation cores or crack tips atomistically. In another publication (Aoki & Pettifor 1991) we present analytic expressions for the response functions beyond the RA and examine in detail the convergence of the bond order series equation (2.8) and the accuracy of linearizing the moment-recursion coefficient relations equations (2.10)–(2.15).

3. Two- and three-body cluster interactions

The linearized bond order potential can also shed light on why some tetragonal tri-aluminides can be stabilized in the cubic form (e.g. Al_3Ti) but others cannot (e.g. Al_3Nb). Carlsson (1989) has recently used the Connolly–Williams method to find the

Figure 5

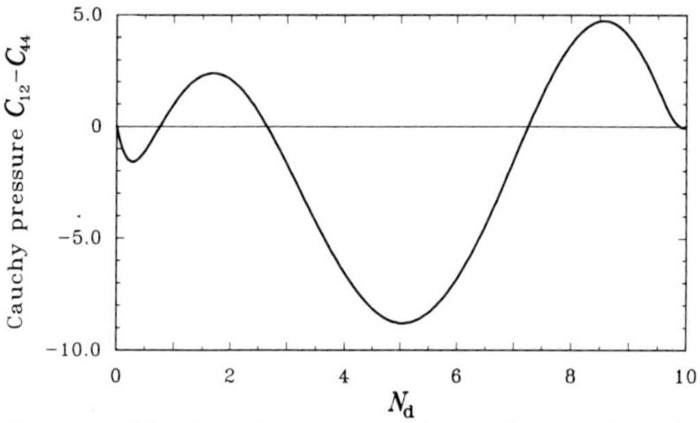

Figure 6

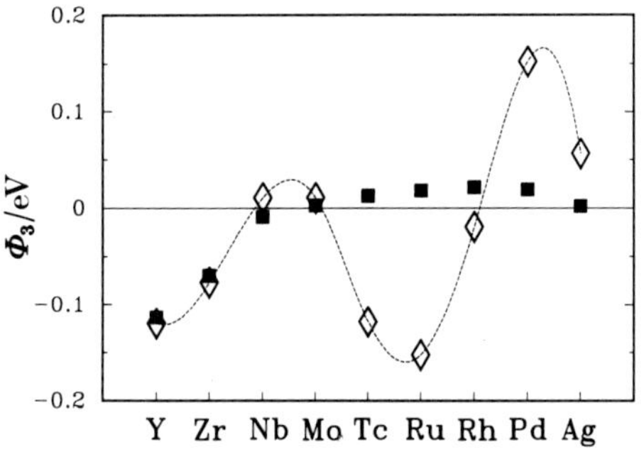

Figure 5. The Cauchy pressure $C_{12}-C_{44}$ as a function of the number of valence d electrons per atom N_d for the BCC lattice.

Figure 6. The three-body cluster interaction Φ_3 for the 4d transition metal aluminides (after Carlsson 1989). The squares denote the predictions of Miedema's model.

effective interaction parameters in a cluster expansion for the total binding energy of FCC-based transition metal aluminides, namely

$$U = \Phi_0 + \Phi_1\langle\sigma_i\rangle + \Phi_2\langle\sigma_i\sigma_j\rangle + \Phi_3\langle\sigma_i\sigma_j\sigma_k\rangle + \Phi_4\langle\sigma_i\sigma_j\sigma_k\sigma_l\rangle. \quad (3.1)$$

Here σ_i is a spin-like variable which takes the value 1 and -1 for transition and aluminium atoms respectively. The averages are taken over all sites, nearest-neighbour pairs, nearest-neighbour triangles, and nearest-neighbour tetrahedra.

The heat of formation for the disordered transition metal aluminide T_cAl_{1-c} is then given by

$$\Delta H^{dis}(c) = -4c(1-c)\,[\Phi_2 - (1-2c)\,\Phi_3 + 2(1-2c+2c^2)\,\Phi_4]. \quad (3.2)$$

We see that regular solution behaviour is determined by the two-body cluster interaction Φ_2, whereas the skewing of the heat of formation curve is determined by the three-body cluster interaction Φ_3. Carlsson's (1989) local density functional calculations predict that Φ_3 oscillates across the transition metal aluminide series as shown for the 4d series in figure 6. For $\Phi_3 < 0$ the parabolic regular solution curve is skewed towards the aluminium rich end, whereas for $\Phi_3 > 0$ the curve is skewed towards the transition metal rich end. Thus figure 6 predicts that FCC-based Al_3Zr (or Al_3Ti) will be more stable than FCC-based $AlZr_3$ (or $AlTi_3$), whereas FCC-based Al_3Nb (or Al_3V) will be less stable than FCC-based $AlNb_3$ (or AlV_3). It is, thus, not surprising that Al_3Ti can be stabilized in the cubic $L1_2$ form but Al_3Nb cannot be. (In the latter case tetragonal distortion lowers the observed DO_{22} lattice by about 1 eV per formula unit compared to the cubic $L1_2$ lattice.)

What is the origin of the oscillations in Φ_3? We see in figure 6 that the Miedema 'macroscopic atom' model does not predict such rapid variations across the series (Carlsson 1989). The beauty of the linearized bond order potential is that it provides for the first time a cluster expansion which is applicable to any lattice and chemical ordering (cf. equations (2.8) and (2.26)–(2.30)).

To understand the quantum mechanical origin of Φ_3 in the transition metal aluminides, let us consider the simpler case of transition metal intermetallics A_cB_{1-c} where the A and B atoms are characterized by atomic d energy levels $-\frac{1}{2}\Delta E$ and $+\frac{1}{2}\Delta E$ respectively and off-diagonal disorder is neglected. The heat of formation may then be written

$$\Delta H^{dis} = \Delta H^{dis}_{vol} + \Delta H^{dis}_{bond}, \quad (3.3)$$

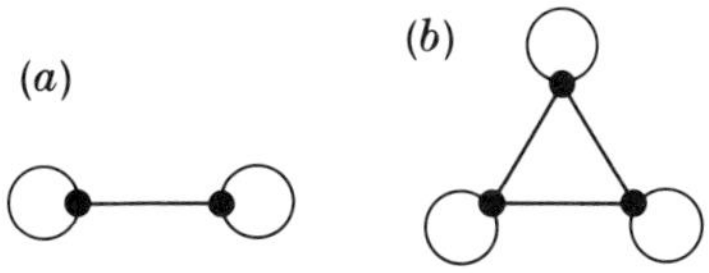

Figure 7. The two-body (*a*) and three-body (*b*) diagrams which contribute to Φ_2 and Φ_3 respectively. The bubbles represent hopping on the same site, which would give a factor $-\frac{1}{2}\Delta E$ $(+\frac{1}{2}\Delta E)$ for an A (B) atom.

where $\Delta H^{\text{dis}}_{\text{vol}}$ is the usual contribution to the heat of formation due to differences in the equilibrium atomic volume of the pure constituents. The contribution from the change in the bond energy may be arranged as

$$\Delta H^{\text{dis}}_{\text{bond}} = \tfrac{1}{2}(10zh)\left\{ \begin{array}{c} 2c(1-c)\,[\Theta^{\text{dis}}_{\text{AB}} - \frac{1}{2}(\Theta^{\text{dis}}_{\text{AA}} + \Theta^{\text{dis}}_{\text{BB}})] \\ +[c(\Theta^{\text{dis}}_{\text{AA}} - \Theta^{\text{A}}_{\text{AA}}) + (1-c)\,(\Theta^{\text{dis}}_{\text{BB}} - \Theta^{\text{B}}_{\text{BB}})] \end{array} \right\}, \tag{3.4}$$

where, for simplicity, the explicit σ, π, and δ bond character has been averaged out in an effective bond integral $h < 0$ (see, for example, p. 80 of Pettifor 1987) and the prefactor 10 accounts for the d band degeneracy. $\Theta^{\text{dis}}_{\text{AB}}$, $\Theta^{\text{dis}}_{\text{AA}}$, and $\Theta^{\text{dis}}_{\text{BB}}$ are the bond orders of the AB, AA, and BB bonds in the disordered A_cB_{1-c} alloy respectively, whereas $\Theta^{\text{A}}_{\text{AA}}$ and $\Theta^{\text{B}}_{\text{BB}}$ are the bond orders of the AA and BB bonds in pure A and pure B respectively.

The response functions describing the AB and the average of the AA and BB bond orders in the disordered alloy are the same, as both are determined by the average of the moments about sites A and B through (2.17) and (2.20)–(2.24). Thus, neglecting the change in the AA and BB bond orders in going from the elemental metals to the alloy, (3.4) for the heat of formation may be written as

$$\Delta H^{\text{dis}}_{\text{bond}} = 20z|h|\,c(1-c)\left\{ \sum_{n=0}^{\infty} \chi^{\text{dis}}_{0n,\,n0}\,\Delta[\delta a_n] + 2\sum_{n=1}^{\infty} \chi^{\text{dis}}_{0(n-1),\,n0}\,\Delta[\delta b_n] \right\}, \tag{3.5}$$

where
$$\Delta[\delta a_n] = [\delta a_n]_{\text{AB}} - [\delta a_n]_{(\text{AA}+\text{BB})/2} \tag{3.6}$$
and
$$\Delta[\delta b_n] = [\delta b_n]_{\text{AB}} - [\delta b_n]_{(\text{AA}+\text{BB})/2}. \tag{3.7}$$

The heat of formation, therefore, depends on those diagrams which are different between the AB and the average AA and BB bond orders. It follows from (2.26)–(2.30) that within the RA the first two nearest-neighbour terms in the cluster expansion are

$$\Delta H^{\text{dis}}_{\text{bond}} = 20z|h|\,c(1-c)\,\{2(h/b)\,(\Delta E/2b)^2\,\hat{\chi}_4[\tfrac{1}{2}(N_{\text{A}}+N_{\text{B}})] + 2(1-2c)\,(h/b)^2\,(\Delta E/2b)^3\,\hat{\chi}_6[\tfrac{1}{2}(N_{\text{A}}+N_{\text{B}})]\}, \tag{3.8}$$

which correspond to the two diagrams shown in figure 7. $\hat{\chi}_4$ and $\hat{\chi}_6$ are functions of the average number of valence electrons per spin per AB bond, namely $\frac{1}{2}(N_{\text{A}}+N_{\text{B}})$.

The two-body and three-body cluster interactions are given by comparing (3.2) and (3.8). Assuming a rectangular density of states of width W for which $|b| = \mu_2^{\frac{1}{2}} = W/\sqrt{12}$ and a close-packed lattice for which $z = 12$, we have

$$\Phi_2 = -5\sqrt{3}\,W(\Delta E/W)^2\,\hat{\chi}_4 \tag{3.9}$$
and
$$\Phi_3 = -\tfrac{5\sqrt{3}}{2}W(\Delta E/W)^3\,\hat{\chi}_6. \tag{3.10}$$

This predicted behaviour of Φ_2 and Φ_3 as a function of the average number of valence electrons per AB bond is similar to that found computationally using the

Connolly–Williams method for transition metal intermetallics (see fig. 1*a* of Sluiter & Turchi 1989) and for the transition metal aluminides in figure 6 (Carlsson 1989). We see, therefore, that the rapid oscillations in Φ_3 are real, reflecting the wave mechanical nature of the three-body diagram in figure 7.

4. Conclusion

We have shown that the theoretically derived bond order potentials should be invaluable for the atomistic simulation of intermetallics where both the angular character and the many-body nature of the potential are important. We illustrated this for the case of metals with negative Cauchy pressure which in the past could not be treated realistically by other available potentials (see, for example, Johnson 1988). In addition we demonstrated that the very important deviations from regular solution behaviour are a direct consequence of an explicit three-body term in a newly derived cluster expansion. It remains to apply these bond order potentials to specific systems such as FCC iridium or $L1_2$ Al_3Ti in order to explore the possible microscopic mechanisms which may be responsible for their unexpected brittleness.

We thank the U.S. Department of Energy, Energy Conversion and Utilization Technologies (ECUT) Materials Program, for financial support under subcontract no. 19X-55992V through Martin Marietta Energy Systems Inc. D.G.P. acknowledges Dr C. T. Liu of the Oak Ridge National Laboratory for his continuing enthusiasm and support for a more fundamental understanding in alloy design.

References

Aoki, K. & Izumi, O. 1979 Improvement in room temperature ductility of the $L1_2$ type intermetallic compound Ni_3Al by boron addition. *Nippon Kinzoku Gakkaishi* **43**, 1190–1195.

Aoki, M. & Pettifor, D. G. 1991 (In preparation.)

Carlsson, A. E. 1989 Cluster interactions and physical properties of Al-transition metal alloys. *Phys. Rev.* B **40**, 912–923.

Dimiduk, D. M. & Miracle, D. B. 1989 Directions in high temperature intermetallics research. *Mater. Res. Soc. Symp. Roc.* **133**, 349–359.

Fu, C. L. 1991 Electronic, elastic and fracture properties of trialuminide alloys: Al_3Sc and Al_3Ti. (In the press.)

Haydock, R., Heine, V. & Kelly, M. J. 1972 Electronic structure based on the local atomic environment for tight-binding bands. *J. Phys.* C **5**, 2845–2858.

Johnson, R. A. 1988 Analytic nearest-neighbour model for fcc metals. *Phys. Rev.* B **37**, 3924–3931.

Nicholson, D. M., Stocks, G. M., Temmerman, W. M., Sterne, P. & Pettifor, D. G. 1989 Structural energy differences in Al_3Ti: the role of tetragonal distortion in APB and twin energies. *Mater. Res. Soc. Symp. Proc.* **133**, 17–22.

Pettifor, D. G. 1983 Electron theory of metals. *Physical metallurgy* (ed. R. W. Cahn & P. Haasen), pp. 73–152. Amsterdam: Elsevier.

Pettifor, D. G. 1987 A quantum-mechanical critique of the Miedema rules for alloy formation. *Solid St. Phys.* **40**, 43–92.

Pettifor, D. G. 1988 Structure maps for pseudobinary and ternary phases. *Mater. Sci. Technol.* **4**, 675–691.

Pettifor, D. G. 1989 New many-body potential for the bond order. *Phys. Rev. Lett.* **63**, 2480–2483.

Pettifor, D. G. 1990 From exact to approximate theory: the Tight Binding Bond model and many body potentials. *Springer Proc. Phys.* **48**, 64–84.

Pettifor, D. G. 1991 It's all very well in practice but what about in theory? *New materials for the next century: a scientific, technological and industrial revolution* (ed. G. A. D. Briggs). Oxford: Blackwells.

Schneibel, J. H. & Porter, W. D. 1989 Microstructure and mechanical properties of $L1_2$-structure alloys based on Al_3Zr. *Mater. Res. Soc. Symp. Proc.* **133**, 335–340.

Sluiter, M. & Turchi, P. 1989 Electronic theory of phase stability in substitutional alloys: a comparison between the Connolly–Williams scheme and the generalized perturbation method. *Alloy phase stability* (ed. G. M. Stocks & A. Gonis), pp. 521–528. Dordrecht: Kluwer.

Subramanian, P. R., Simmons, J. P., Mendiratta, M. G. & Dimiduk, D. M. 1989 Effect of solutes on phase stability in Al_3Nb. *Mater. Res. Soc. Symp. Proc.* **133**, 51–56.

Discussion

J. N. MURRELL (*University of Sussex, U.K.*). The concept of bond order is not sufficiently well defined outside the simplest systems (e.g. organic hydrocarbons) for this to form the basis of quantitatively successful theories. The reason for this is that bond orders for, say, triangles of different side lengths, are not unambiguously defined.

D. G. PETTIFOR. The simple two-centre, orthogonal tight binding Hückel theory has been shown to predict successfully the trends in crystal structure of elemental systems through the periodic table, of binary intermetallics through the AB, AB_2, and AB_3 structure maps, and of small microclusters of carbon and silicon. The many-body potential for the bond order is derived directly from this TB model and, therefore, is well based to predict trends in structural behaviour.

A. COTTRELL (*Department of Materials Science, Cambridge, U.K.*). To avoid brittleness I think a short Burgers vector is needed, i.e. not a superlattice vector. This means that the two atoms in the intermetallic have to be fairly indifferent towards one another, which implies a modest melting point. That from the practical point of view is a depressing conclusion, but it might be possible to bypass it with a disordered intermetallic. Does Professor Pettifor have a view on this?

D. G. PETTIFOR. My only comment is that the well-known ductilization of Ni_3Al polycrystals by adding boron appears to be due to the boron attracting excess nickel to the grain boundary, thereby setting up a disordered layer which improves the mechanical properties. We have still to perform atomistic simulations of crack tips in elemental and binary metals with these new angularly dependent bond order potentials, and to study the influence of ordering energy on the mechanical response.

A. M. STONEHAM (*Harwell, U.K.*). Professor Pettifor has used a non-self-consistent approach; for metals, one can see this could work, but for semiconductors (where defect charge state matters, e.g. in Fermi level effects on dislocation motion) surely more is required?

D. G. PETTIFOR. Yes, of course.

Quantum percolation theory and high-temperature superconductivity

By J. C. Phillips

AT&T Bell Laboratories, Murray Hill, New Jersey 07974, U.S.A.

An overall view of the methodology of quantum percolation theory as applied to high-temperature superconductivity clarifies the differences in its approach compared to others. Several crucial predictions have been confirmed by experiment, including the percolative nature of the superconductive transition itself.

1. Introduction

Many-layered cuprates (such as $YBa_2Cu_3O_7$) are superconductive with transition temperatures about five times higher than those obtainable in intermetallic compounds such as Nb_3Sn. These materials are often insulating when initially formed and become metallic and superconductive only when subjected to carefully controlled oxidation or reduction. Even then their room-temperature resistivities are usually larger (greater than 200 μΩ cm) than the Mooji resistivity (*ca.* 100 μΩ cm) which separates ordered 'good' metals ($d\rho/dT > 0$, with predominant scattering by phonons) from disordered 'bad' metals ($d\rho/dT < 0$, residual scattering by disorder). We may say that the highest transition temperatures are found not in 'good' metals like Al and Pb, but in oxides which are 'bad' metals, with resistivities similar to those found in metallic glasses.

Most theoretical efforts to explain this unexpected situation have focused on the search for alternatives to the electron–phonon interaction which is the basis for the BCS theory of superconductivity in intermetallic compounds (Phillips 1989*a*). The two alternatives discussed most often are electron–magnon and electron–exciton interactions. These alternatives are interesting (and perhaps even refreshing), but there are extremely persuasive reasons for excluding them at present. The strongest is probably Ockham's razor, which says that however attractive novelty may be for its own sake, we should insist that the standard electron–phonon mechanism be disproven before we discard it. The efforts to do this so far have relied mainly on naive interpretations of data from heterophase samples (where magnetism and superconductivity coexist, but in different parts of the sample) or on misassignment of infrared absorption bands (due not to excitons but to defects) or to misinterpretations of the isotope effect. Meanwhile, there is growing evidence (Phillips 1990*a*; Reichardt *et al.* 1989; Franck *et al.* 1991) for strong electron–phonon interactions from isotope effect experiments and from phonon softening at T_c. Finally there is the generic assessment (Phillips 1989*a*) which shows that one-electron–ion interactions are 10–20 times stronger than electron–magnon or electron–exciton interactions, so that the latter could never be taken seriously as a mechanism for Cooper pairing in high T_c materials.

Phil. Trans. R. Soc. Lond. A (1991) **334**, 451–458

Printed in Great Britain

2. Quantum percolation theory (QPT)

The alternative to exotic interactions is percolation, or two-component Fermi liquid theory, which says that near a metal–insulator transition the electronic states near the Fermi energy E_F can in favourable circumstances be separated into two groups, localized and extended. The electron–phonon interaction, although not large on the average, can be large for a few extended states because it is small for many localized states.

The answer to the key question of whether Fermi energy extended and localized states can coexist and be separated into a two-component Fermi liquid in a disordered metal or in the impurity band of a semiconductor depends on dimensionality d. For $d < d_m = 2$ all states are localized, whereas for $d > d_m$ some extended states may exist. Because the marginal dimensionality $d_m = 2$, it is natural to suppose that this is the reason that the layered cuprates are both bad normal metals and high-temperature superconductivity (HTS). The derivation of the marginal dimensionality $d_m = 2$ is topological (orbital) and does not depend on the nature of the electron–electron interactions (Coulomb, exchange, excitonic, or what have you), emphasizing the same point that exotic interactions are not the key to HTS.

One of the peculiarities of localization theory is that there seem to be as many theories as there are theorists. Considering that classical percolation, a simpler situation, is always treated by numerical rather than analytic methods, it is not surprising that quantum percolation is a complex subject. Most of the final decisions must be made by experiment, but the possibilities can be listed theoretically. Some of the central issues are matters of style, and here consistency is important. For example, the main ideas of QPT have not changed since I first presented them (Phillips 1983*a*, *b*, 1988), and several predictions have been successful. This has not been the case for any of the theories based on exotic interactions.

The main ideas of quantum percolation theory, in addition to the separability of localized and extended states for $d > d_m = 2$, in the context of the cuprates, are the following. We are concerned with the states contained in the energy shell $|E - E_F| \lesssim E_c$, where $E_c \approx 0.1 - 0.2$ eV is only somewhat larger than the maximum vibrational energy (*ca.* 0.05 eV). In addition to the planar CuO_2 states near E_F (which would be metallic if they were three dimensional) there are defect states with energy E_d in the semiconductive layers (such as BaO in $YBa_2Cu_3O_7$) at $|E_d - E_F| \lesssim E_c$. These defect states not only pin E_F but they also act as electrical bridges between CuO_2 planes or CuO chains. Conduction in these planes is blocked by domain walls generated by residual disorder, with a typical domain wall spacing $w \approx 100$ Å†. These electrical bridges generate an effective dimensionality $d = 2 + \epsilon$ associated with percolative paths of the type shown in figure 1.

Elucidation of the nature of localized and extended states is provided by spectroscopic experiments on $YBa_2Cu_3O_7$, notably Raman scattering (Cooper *et al.* 1988; Phillips 1989*b*) and single-domain polarized reflectivity studies (Schlesinger *et al.* 1990) of the electronic continuum near E_F both above and below T_c. For $T > T_c$ the continuum observed by Raman scattering for $\hbar\omega \lesssim E_c$ is somewhat enhanced near $\omega = 0$ by defects (probably oxygen vacancies $O^{\square}$ on CuO chains) but it is approximately constant, which suggests that the measured scattering strength

† 1 Å = 10^{-10} m.

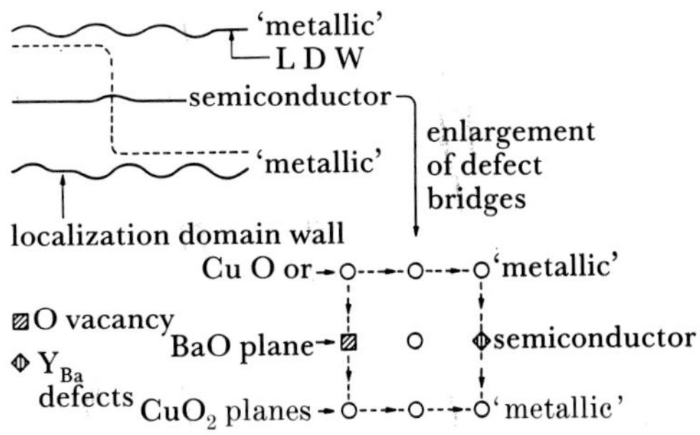

Figure 1. A sketch of interplanar current paths avoiding localization domain walls by utilizing Fermi-energy defects.

provides a faithful representation of $N(E)$. For $T \ll T_c$ this same continuum scattering is proportional to $|E-E_F|$, and in spite of superposed vibrational bands, a planar energy gap E_g is recognizable near 500 $cm^{-1} \approx 8\ kT_c$. Two-component Fermi liquid theory (FLII) readily explains the observed spectral changes. For $T > T_c$ both extended and localized states contribute to the scattering. Well below T_c, according to FLII, only the extended Fermi liquid states have become superconductive and been pushed above E_g. The localized Fermi liquid states are confined to incoherent planar domains and do not become superconductive. Their density of states $N_l(E_F)$ is proportional to $|E-E_F|$ because of maximal elastic metallic hybridization (Phillips 1990*b*).

FLII immediately makes crucial predictions concerning the frequency and/or temperature dependence of the Drude currents carried by extended states. The interplanar coherence of the extended states is fragile and is destroyed by elastic scattering by residual disorder. In one-component Fermi liquids (FLI) in strongly disordered metals elastic scattering makes the residual resistivity nearly constant (within a few per cent). In FLII we have a selection rule for elastic scattering of extended states: they are always scattered into localized states. But while $N_e(E)$, the density of extended states, is nearly constant near E_F (as in a normal Fermi liquid), $N_l(E) \propto |E-E_F|$. This means that $\Gamma_e(\omega, T)$, the scattering rate for extended states, is proportional to (ω, T), in agreement with experiment (Schlesinger *et al.* 1990). Notice that these predictions do not require any modification of the extended state Fermi surface states, whereas some other attempts using FLI to explain $\rho(\omega,T)$ require (Varma *et al.* 1989; Anderson 1990) that the renormalized equasi-particle strength $z(E_F) = 0$.

It appears then that the crucial concept which is pivotal to understanding the electronic structure of HTS is QPT or FLII. This concept originally arose (Phillips 1983*a*, *b*, 1988) in an effort to understand the metal–semiconductor transition in impurity bands (Si:P). In earlier work Mott assumed (Mott 1972; Cohen 1970) that localized and extended states could not coexist, and therefore that this transition was always discontinuous, whereas Anderson and coworkers showed (Abrahams *et al.* 1979) with a classical model that it could be continuous, and that in some cases the marginal dimensionality for achieving metallic conductivity might be $d = d_m = 2$. The key step which I took (Phillips 1983*a*, *b*, 1988) was to argue that not only could localized and extended states coexist, but that also in certain cases these coexisting states could be separated. This separation leads to the two-component Fermi liquid model (FLII) which differs drastically from the ordinary Fermi liquid model FLI which is used to describe the normal state in conventional metallic superconductors. Although classical scaling arguments (Abrahams *et al.* 1979) were used only to guess that $d_m = 2$, I was able to derive this result from the relation $d_m - 1 = \frac{1}{2}d_m$. This relation is the quantum analogue, based on the uncertainty principle, of a similar relation derived in classical random-field Ising models (Imry & Ma 1975).

3. Some general remarks

The separation of localized and extended states in a highly disordered medium is possible only as a hypothetical process, because near the metal–insulator transition the 'signal' (the extended states) is very small compared to the 'noise' (the localized states). The possibility of such a separation varies from case to case. It cannot be proved, but must actually be decided by experiment. For instance, this separation is possible in Si:P, but only for uncompensated samples (Phillips 1983*a*, *b*, 1988). In axiomatic set theory (Cohen 1966; Tiles 1989) one can show that once extended states exist, they can be separated from localized states regardless of energy. (This is called the subset axiom.) The separability is a general property, and it does not require us to find a constructive algorithm, that is to say the correct unitary transformation which actually performs the separation. (If we could find such an algorithm, we would validate the axiom of choice (AC) for our disordered system. But the remaining axioms, including the subset or separability axiom, are valid independently of the AC, according to Gödel and Cohen (Cohen 1966; Tiles 1989). Such an AC algorithm is especially difficult to construct in the face of the uncertainty principle, but its absence in no way impairs the validity of the subset axiom. The AC is important if we want to impose internal order on (or alphabetize) the subset of extended states, for example, but this is not necessary in the present circumstances, where we need to know only $N_l(E)$ and $N_e(E)$.) In the case of HTS, it is just the high T_cs, together with the layered structures ($d \approx d_m$) that convinced me that a separation is possible.

At this point the reader may well wonder why such a simple construct as FLII has not already gained general acceptance. If we discount some of the obvious (but practically important) explanations, such as a deluge of theories of HTS, then we may hazard the following observations. QPT has not been generally accepted (Lakner & Löhneysen 1990) even as an explanation (Phillips 1983*a*, *b*, 1988) for the conductivity exponent m in the relation $\sigma \propto (n-n_c)^m$ in Si:P, where $m = 0.50$ in the uncompensated state, even though it is the only theory that explains the result. (All other theories give $m \geqslant \frac{2}{3}$.) Even in this more leisurely field, the separability of localized and extended states is not generally accepted (Lakner & Löhneysen 1990). Probably this comes from the reluctance of most physicists to accept as a probability the existence of the abstract unitary transformation which separates $\{l\}$ and $\{e\}$ without explicit examples. (However, when such an abstract separation into localized domains and domain walls is made in the classical random-field Ising model (Imry & Ma 1975), it is considered obvious and seminal. Mathematically the two situations are equivalent, but of course the separation is easier to visualize classically in real space than quantum mechanically in Hilbert space.)

We can understand the difference in philosophy between FLI and FLII in another way. In FLI the effects of scattering by disorder on itinerant wave functions are treated by perturbation theory, either in the one-electron approximation or by using renormalization techniques to describe many-electron interactions. What many scientists apparently do not realize is that renormalization methods, being perturbative, are limited to the range $|1-z| \ll 1$ just like ordinary one-electron theory. The real utility of FLI lies thus in removing divergences associated with strong electron–electron long-range Coulomb repulsion, and it brings very little that is new to our understanding of the effects of short-range disorder near the metal–insulator transition where the renormalization factor $z \ll 1$.

Excellent illustrations of the difference between the set-theoretic and perturbative approaches are provided by numerical studies (Feng *et al.* 1985) of the stiffness transition of the vibrations of amorphous or glassy systems. The $N \times N$ dynamical matrices describing these systems are very complex, with *ca.* $\frac{1}{2}(1-z)\,N^2$ off-diagonal matrix elements. If these systems were treated in the spirit of FLI, we would construct a virtual crystal model, obtain its eigenvalues $\{\omega^2\}$, and then scale these eigenvalues by $\omega_n^2 + (1-z)\omega_{\text{od}}^2$, where ω_n^2 are the diagonal matrix elements and ω_{od}^2 represents off-diagonal effects. This gives good answers for z near 1, but near the percolation threshold $z \ll 1$ the answers are qualitatively wrong. In particular, perturbation theory will never yield the cyclical (Goldstone) modes which are found by exact matrix diagonalization. Thus the recent failures (Varma *et al.* 1989; Anderson 1990) of FLI to describe HTS are not technical in origin, and they cannot be repaired by technical refinements. The failures are conceptually intrinsic to FLI and cannot be overcome without replacing FLI by FLII.

A deeper and more fundamental objection to separability is that in any two-fluid model the entropy is lower than in a one-fluid model. By assuming separability but not actually displaying it, are we not in effect postulating a Schrödinger demon who could violate the second law of thermodynamics, which in other contexts leads to perpetual motion machines? The answer is no. Separability is important only to transport (including optical) properties and not to thermal experiments. In transport theories at some stage a relaxation time approximation is always made. This approximation is applied to diagonal elements of the density matrix and it yields different results for different basis sets. It is therefore essential that we choose the correct basis states before we discuss any transport property, including superconductivity.

I conclude by mentioning that in a recent experiment on single-crystal YBCO, Palstra *et al.* (1990) have shown that there is substantial entropy transport in crossed electronic and magnetic fields above T_c. This result can also be explained by QPT, and it provides an excellent illustration of the argument made above that FLII is internally consistent (J. C. Phillips, unpublished work).

References

Abrahams, E., Anderson, P. W., Licciardello, D. C. & Ramakrishnam, T. V. 1979 Scaling theory of localization: absence of quantum diffusion in two dimensions. *Phys. Rev. Lett.* **42**, 673.

Anderson, P. W. 1990 'Luttinger-liquid' behavior of the normal metallic state of the 2D Hubbard model. *Phys. Rev. Lett.* **64**, 1839.

Cohen, P. J. 1966 *Set theory and the continuum hypothesis.* New York: Benjamin.

Cohen, M. H. 1970 Review of the theory of amorphous semiconductors. *J. non-crys. Solids* **4**, 391.

Cooper, S. L., *et al.* 1988 Raman scattering from superconducting gap excitations in single crystal $YBa_2Cu_3O_7$. *Phys. Rev* B **37**, 5920.

Feng, S., Thorpe, M. F. & Garboczi, E. 1985 Effective-medium theory of percolation on central-force elastic network. *Phys. Rev.* B **31**, 276.

Franck, J. P., Jung, J. & Mohamed, M. A.-K. 1991 *Proc. XIX Int. Conf. Low Temp. Phys.* (In the press.)

Imry, Y. & Ma, S.-K. 1975 Random-field instability of the ordered state of continuous symmetry. *Phys. Rev. Lett.* **35**, 1399.

Lakner, M. & Löhneysen, H. V. 1990 Localized versus itinerant electrons at the metal–insulator transition in Si:P. *Phys. Rev. Lett.* **64**, 108.

Mott, N. F. 1972 Conduction in non-crystalline systems. *Phil. Mag.* **26**, 1015.

Palstra, T. T. M., Batlogg, B., Schneemeyer, L. F. & Waszcak, J. V. 1990 Transport entropy of vorton motion in $YBa_2Cu_3O_7$. *Phys. Rev. Lett.* **64**, 3090.

Phillips, J. C. 1983*a* Why localized and extended states can coexist and be separated. *Solid St. Commun.* **47**, 191.

Phillips, J. C. 1983*b* Separable model of the metal–insulator transition in Si:P. *Phil. Mag.* B **47**, 407.

Phillips, J. C. 1988 Non-local theory of the metal–insulator transition in Si:P. *Phil. Mag.* B **58**, 361.

Phillips, J. C. 1989*a* *Physics of high-T_c superconductors*. Boston: Academic.

Phillips, J. C. 1989*b* *Phys. Rev.* B **40**, 7348.

Phillips, J. C. 1990*a* *Phys. Rev. Lett.* **64**, 1605.

Phillips, J. C. 1990*b* Phase-space restrictions on localized states in percolative metals. *Solid St. Commun.* **73**, 135.

Reichardt, W. *et al.* 1989 *Physica* C **162**, 464.

Schlesinger, Z. *et al.* 1990 Superconducting energy gap and normal-state conductivity of a single-domain $YBa_2Cu_3O_7$ crystal. *Phys. Rev. Lett.* **65**, 801.

Tiles, M. 1989 *The philosophy of set theory*, pp. 121–134. Oxford: Blackwell.

Varma, C. M. *et al.* 1989 Phenomenology of the normal state of Cu–O high temperature superconductors. *Phys. Rev. Lett.* **63**, 1996.

Discussion

A. O. E. Animalu (*University of Nigeria, Nsukka, Nigeria*). Would it not be possible and even natural to replace the localized states by a narrow Cu(3d) band that crosses and hybridizes with an extended broad (2p) band, in a two-band model; in which case there may be a connection between the QPT and a two-band model?

J. C. Phillips. This thought also occurred to me, but I finally abandoned it because in spatially homogeneous two-band models both bands become superconductive at a common transition temperature. In the cuprates, however, the states in the gap remain there down to $T \ll T_c$. It therefore seems natural to assume that these states are localized and that for them $\lambda_{ep} - \mu^* < 0$, where λ_{ep} is the electron–phonon coupling strength and μ^* is the renormalized Coulomb repulsion.

P. W. Anderson (*Princeton University, U.S.A.*). (1) The isotope effect was discussed by Fisher *et al.* They remarked that there is a relatively large contribution coming from the 'natural' effect of changes in hopping matrix elements, which are exponential in distance. These give us little confidence in deductions from the isotope effect.

(2) I am not aware of serious controversy in the theory of localization. It is accepted that the Si:P case is an exceptional one of strong interactions which has been explained in most features by Kotliar, di Castro and others. Localization in percolating networks was well treated by B. Shapiro and has no particular unusual

features. In no case can localized and extended régimes overlap near the Fermi energy.

(3) It is often claimed that the cuprates are poor conductors. This is far from true: their room-temperature conductivity per electron is only 2–3 less than pure copper. In particular, localization phenomena always occur close to the Mott–Yoff–Regal criterion $k_F\, l \approx 1$ while their number is greater than 10 in these metals.

J. C. PHILLIPS. The paper by Fisher *et al.* was never taken seriously by me for several reasons, including fudging of several parameters. In any event its relevance disappeared after large isotope shifts were found in LSCO, as I predicted, and more recently in (Y, Pr) BCO alloys, as discussed in the reply to Professor Salje.

The wide variety of viewpoints on localization is a characteristic aspect of the subject which is discussed extensively by Sir Nevill Mott in almost all of his excellent papers on this subject during the 1970s and 80s. He, himself, for example, adhered to the discontinuous model of the Si:P transition until quite recently, when he seems to have tacitly adopted the FLII model. The various scaling papers mentioned by Professor Anderson, curiously enough, seem to have regressed to the discontinuous model, a large step backwards from his own continuous (but linear rather than square root) scaling model of 1979. I have recently shown that strong interactions not only do not explain the square root dependence of σ on $(n - n_c)$ in the $T = 0$ limit, but they are also not needed to understand the cross-over from square-root to linear behaviour in partly compensated samples. Also, I have shown that strong interactions, as treated by first-order perturbation theory by Altschuler and Aronov, are not the appropriate way to explain $\sigma = \sigma(T) - \sigma(0)$ for n near n_c, but that this difference can be explained entirely by elementary fluctuation theory in the context of quantum percolation. Of course, in the end 'forces are at work', as someone said, and it is the competition between Coulomb repulsion and kinetic energy that is responsible for metal–semiconductor transitions in both Si:P and high T_c cuprates. However, this competition takes place in highly disordered media, and it is the disorder itself, and not the strength of the forces, which governs the functional behaviour in the critical region in Si:P and which generates the giant electron–phonon couplings in the cuprates. As for quantum percolation itself, Shapiro and others discuss only classical percolation. Quantum percolation, because of its two-component nature, is fundamentally different from classical percolation.

As for whether the cuprates are good conductors or not, experimentalists are well aware that the as-prepared samples are generally green, and that subtle oxidation or reduction procedures are required to produce materials which are 'good' metals in Professor Anderson's sense. However, even then, because of marginal dimensionality, a mean free path of order 50–100 Å in a cuprate plane is no guarantee that we have a 'good' metal, because electrons in the absence of interplanar defect bridges will still be localized in planar domains 100 Å in diameter. It is the density of these bridges that explains why the number of effective carriers is small and why the associated mean free paths are long. All of these points are examined at length in my book.

E. SALJE (*Cambridge University, U.K.*). The arguments of Professor Phillips seem to be supported by recent experimental results. (1) The isotope effect in YBCO is finite. (2) Phonon renormalization during the superconducting phase transition is as large as 3 cm^{-1} in YBCO. (3) The corrected carrier density during condensation is *ca.* 10%

of the total carrier condensation. However, does not Co doping with increased O concentration change T_c for small doping levels?

J. C. Phillips. With regards to the isotope effect, recent experiments by Franck *et al.* (1991) on the isotope shift in (Y, Pr)BCO alloys confirm that its small value in YBCO is accidental and is probably the result of anharmonic variations in domain wall width (Phillips 1990*a*), where I also discussed phonon renormalization. The reduction in the effective carrier density that he mentions is fully consistent with interlayer percolation via apical oxygen vacancy states, with the apical oxygen vacancy concentration being about 6%, as revealed by recent well-refined diffraction data on an untwinned YBCO sample. Finally, I have already discussed the effect of Co, Ni and Zn doping on T_c (Phillips 1989*a*). There I concluded that serious questions arise concerning dopant homogeneity on the appropriate length scale, which is less than 100 Å. More recent data appear to indicate better homogeneity, but it is still doubtful whether adequate levels of homogeneity have been achieved.

Electronic structure of the high T_c superconductors

By T. M. Rice[1], F. Mila[2] and F. C. Zhang[3]

[1] *Theoretische Physik, Eidg. Tech. Hochschule – Zürich 8093 – Zürich, Switzerland*
[2] *Serin Physics Laboratory, Rutgers University, P.O. Box 849, Piscataway, New Jersey 08854, U.S.A.*
[3] *Department of Physics, University of Cincinnati, Cincinnati, Ohio 45221, U.S.A.*

The high T_c superconductors have one structural element in common, namely CuO_2 planes which are lightly doped away from an average valence of Cu^{2+}. In the absence of this doping the planes are in a Mott insulating state with local $S = \frac{1}{2}$ moments on each Cu-site. There is considerable evidence both experimental and theoretical supporting the assignment of the extra holes, introduced by doping, to the antibonding O 2p-orbitals. The strong hybridization between these orbitals and the central Cu 3d-orbitals makes it favourable to bind the extra hole with a Cu^{2+} local moment to form a spin singlet state centred on a CuO_4 square. This singlet, however, is mobile and the combination of mobile charged singlets and local spins is described by the so-called *t–J* model. This has a number of consequences which can be tested experimentally. For example, one can use this model to estimate the hyperfine coupling constants which are measured in NMR experiments. The prediction that the only spin degrees of freedom are the local Cu^{2+} spins even upon doping can also be tested in NMR experiments.

1. Introduction

The discovery of high-temperature superconductivity in the cuprate oxides (Bednorz & Müller 1986) has been without doubt the most exciting and unexpected development in solid state physics in the past decade. Further this discovery poses the most urgent problem to the theory of solids. What is so special about this small group of compounds that causes them to be superconducting at temperatures *ca.* 10^2 K? Naturally to begin to answer this question we need to understand the relevant electronic structure at low energies. This in turn is determined by the bonding pattern and the crystal structure.

The cuprate materials are complex materials with large unit cells but they have a single common active element namely CuO_2 planes with four-fold coordinate Cu atoms and two-fold coordinated O atoms. When these materials are prepared so that the formal Cu-valence is 2+ and O-valence is 2–, the CuO_2 are magnetically ordered (Mott) insulators at low temperatures. Superconductivity arises when the counter ions are changed slightly so as to raise the formal valence of the copper ions to $Cu^{2+\delta+}$ with $\delta \approx 0.2$. This change is accompanied by only slight changes in the Cu–O bond lengths and no essential change in the Cu–O bonding pattern as determined by the crystal structure. This tells us right away that the electronic structure must evolve in a continuous way away from the Mott insulating state. Any real change in the electronic structure requires a change in the bonding pattern which in turn must show up in the crystal structure. Yet the crystal structure of the key CuO_2 plane evolves slowly and continuously. This fact alone argues that we are dealing with a doped Mott insulator.

Right at the outset of this problem, this idea was used by Anderson (1987*a*, *b*) to propose that the solution to the high T_c problem lies in the study of strongly correlated electrons near the Mott insulating state and that the simplest model of this case, namely a one-band model, would suffice to understand the physics of this type of quantum fluid. In spite of a tremendous effort in the past years our understanding of this problem is far from complete (e.g. see the recent Los Alamos conference (Bedell *et al.* 1990)). This talk will concentrate on the simpler questions concerning the reduction of this complex electronic system to the one-band model. First we review the line of reasoning connecting the larger energy scale and larger electronic model to the reduced one-band model at the low energy scale. Secondly a comparison can be drawn with the predictions of this model and various experiments. Nuclear resonance techniques are powerful probes of electronic structure on the low energy scale and give us sensitive tests on the low energy scale of the assumptions in the one-band model. Further tests can be made using high energy spectroscopy particularly, X-ray absorption and electron energy loss spectroscopies.

2. Electronic structure of the CuO_2 planes

It is generally accepted that a good starting point to describe the CuO_2 planes is a multiband model composed of $3d_{x^2-y^2}$ orbitals on the Cu-sites and $2p_{x,y}$ orbitals on the O sites (Emery 1987). This gives a model hamiltonian with no fewer than five orbitals per unit cell and when we include the various interatomic (U_{pd}) and intraatomic (U_d, U_{pp}) Coulomb forces in each unit cell we arrive at the following hamiltonian (in hole notation relative to a filled shell ($3d^{10}, 2p^6$) configuration)

$$\begin{aligned} H = &\sum_{i\sigma} \epsilon_d \, d^+_{i\sigma} d_{i\sigma} + \sum_{lm\sigma} \epsilon_p \, p^+_{lm\sigma} p_{lm\sigma} + \sum_{\langle i;lm\rangle} t_{pd\sigma} \, p^+_{lm\sigma} d_{i\sigma} + \text{h.c.} \\ &+ \sum_{\langle l,m;l',m'\rangle} t_{pp} \, p^+_{lm\sigma} p_{l'm'\sigma} + \text{h.c.} + \sum_i U_d \, n_{i\uparrow} n_{i\downarrow} \\ &+ \sum_{\substack{\langle i;lm\rangle \\ \sigma\sigma'}} U_{pd} \, n_{lm\sigma} n_{i\sigma'} + \sum_l \sum_{m\sigma \neq m'\sigma'} U_{pp} \, n_{lm\sigma} n_{lm'\sigma'} \end{aligned} \tag{1}$$

with $d^+_{i\sigma}$ as the creation operator for a hole in a $3d_{x^2-y^2}$ state on Cu, $p^+_{lm\sigma}$ for a hole in a $2p_{x(y)}$ state on O, etc.

This is a very complicated hamiltonian with many parameters. Depending on the values of these parameters, different limiting behaviours can occur at low energies. Various methods have been used to estimate the values of the parameters that enter equation (1). Because the Mott insulating state cannot be described by direct application of density functional theory, new techniques have been devised to make use of the strengths of the density functional method in estimating total energies and densities and from them to obtain parameter values. The values obtained in this way (see, for example, Hybertsen *et al.* 1989; McMahon *et al.* 1989) are typical and are quoted in table 1. These values also agree with early empirical estimates by Mila (1988) and a recent more extensive analysis by Eskes & Sawatzky (1990) (see table 1).

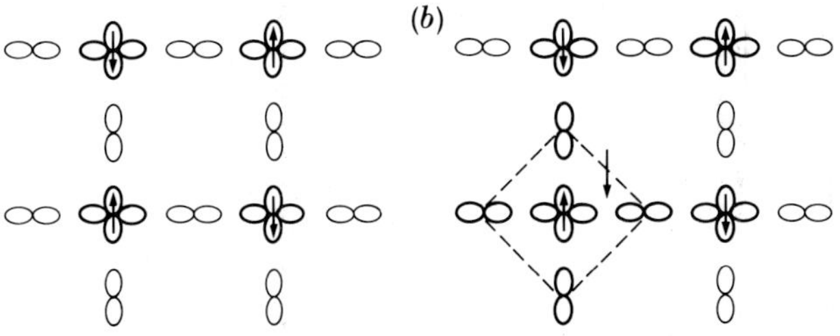

Figure 1. A schematic representation of electronic structure of undoped (*a*) and the hole doped (*b*) CuO_2 plane.

Table 1. *Parameters of the multiband hamiltonian* (1)

(All energies are in electronvolts. The energy parameters in the multiband hamiltonian (1) obtained from constrained local density calculations (first column) and empirical values (second column).)

	Hybertsen *et al.* (1989)	Eskes & Sawatzky (1990)
$\epsilon_p - \epsilon_d$	3.6	2.75–3.75
$t_{pd\sigma}$	1.5	1.5
t_{pp}	0.65	0.65
U_d	10.5	8.8
U_d	4	6
U_{pd}	1.2	< 1

If we look at table 1, the largest Coulomb interaction is the on-site interaction between d-electrons on Cu sites, U_d, and the largest hopping integral is that for the σ-bond between a $3d_{x^2-y^2}$ Cu orbital and a neighbouring $2p_{x(y)}$ O orbital, $t_{pd\sigma}$. If we simplify to these terms only, i.e. set $t_{pp} = U_p = U_{pd} = 0$ and notice that $t_{pd\sigma}/(\epsilon_p - \epsilon_d)$ ($\approx \frac{1}{3}$) is rather small, we can make a perturbation theory in this parameter (Zhang & Rice 1988). First for the case of exactly 1 hole per Cu, i.e. a formal valence of Cu^{2+}, then this hole will sit primarily in a $3d_{x^2-y^2}$ orbital and have some hybridization with the neighbouring 2pO orbitals through the $t_{pd\sigma}$ term. Note it is this strong hybridization that determines the CuO_2 planar structure. It is straightforward to estimate the Heisenberg nearest neighbour (NN) exchange coupling using Anderson superexchange theory. The result in perturbation theory is

$$J = 4t^4_{pd\sigma}/(\epsilon_p - \epsilon_d)^2 U + 4t^4_{pd\sigma}/(\epsilon_p - \epsilon_d)^3. \tag{2}$$

In this case the low energy hamiltonian is simply a Heisenberg spin hamiltonian. There is a clear separation between charge excitations which have a gap and the spin excitations described by the Heisenberg term. Recently Hybertsen *et al.* (1990) went further and by fitting to results of an exact diagonalization of the hamiltonian (1) on small clusters to a Heisenberg model, they could extract a value of J including all terms in equation (1). Their value of $J = 0.13$ eV agrees well with experimental estimates. The electronic configuration is sketched in figure 1*a*.

The corrections to an NN Heisenberg hamiltonian are estimated to be small. For example, the next nearest neighbour term is estimated to be only of order 10^{-2} J. Similarly the Dzyaloshinskii–Moriya interaction which occurs in the presence of slight distortions away from the ideal CuO_2 planar structure is of this order of magnitude too (Coffey *et al.* 1990). Schmidt & Kuramoto (1990) have suggested a large four-site term but this suggestion has lacked experimental confirmation up till now.

While the description of the Mott insulator is more or less agreed upon, there has been much more controversy about the effect of doping. It is clear, however, that, with parameter $U_\mathrm{d} > \epsilon_\mathrm{p} - \epsilon_\mathrm{d}$, the added holes in the hole doped materials must go primarily into O orbitals. The question then is whether this will require a multiband description such as the multiband hamiltonian (1) proposed by Emery (1987), or whether it can still be reduced to single band hamiltonian as Anderson proposed (1987*a*, *b*).

If we start with a simplified version of (1), retaining only the U_d and $t_{\mathrm{pd}\sigma}$ terms as the most important ones, then progress can be made by treating the ratio $t_{\mathrm{pd}\sigma}/(\epsilon_\mathrm{p} - \epsilon_\mathrm{d})$ as a small parameter in perturbation theory. First we consider a single Cu-ion with its four O-neighbours. When we place two holes on this square complex and seek to maximize the gain from the hybridization energy, $t_{\mathrm{pd}\sigma}$, it is immediately clear in second-order perturbation theory, that this is achieved by placing the second hole in a combination of the O p-orbitals with the same symmetry as the central $3\mathrm{d}_{x^2-y^2}$ orbital and also by forming a spin singlet. This is illustrated in figure 1*b*. This state as shown by Zhang & Rice (1988), obtains a binding energy relative to the non-bonding state of

$$E_{\text{singlet}} \approx -8t^2_{\mathrm{dp}\sigma}\left(\frac{1}{\epsilon_\mathrm{p}-\epsilon_\mathrm{d}}+\frac{1}{U+\epsilon_\mathrm{d}-\epsilon_\mathrm{p}}\right). \tag{3}$$

This strong binding suggests immediately that the low energy properties will be governed by these singlets. Note, however, that the singlet state on one $\mathrm{CuO_4}$ has a considerable overlap with that on neighbouring squares (there is a O-site common to both) so that there will be a substantial NN hopping term (*ca.* $-1.5t^2/(\epsilon_\mathrm{p}-\epsilon_\mathrm{d})$) in perturbation theory. In the dilute limit of hole doping the charge will be carried by the tightly bound singlets moving in the $S = \frac{1}{2}$ background which in turn are coupled by a Heisenberg term. This gives us the t–J hamiltonian

$$H_{t-J} = \sum_{\langle ij\rangle\sigma} t_{ij}(1-n_{i-\sigma})\,\mathrm{d}^{+}_{i\sigma}\,\mathrm{d}_{j\sigma}(1-n_{j-\sigma}) + \mathrm{h.c.} + J\sum_{\langle ij\rangle} \boldsymbol{S}_i\cdot\boldsymbol{S}_j, \tag{4}$$

where the sum over i, is restricted to Cu sites. This is the same form as one obtains from a one-band Hubbard model in the case of dilute doping away from half-filling and in the limit $t/U \rightarrow 0$. In this limit of the Hubbard model the ratio $J/t \ll 1$, but in the present case one should treat this ratio as a material parameter, to be determined.

A similar result was obtained by Eskes & Sawatzky (1988). They treated the hamiltonian (1) as having the periodic Anderson model form with the d-states as the strongly correlated atomic states and the p-states as forming the equivalent of the conduction band states. They solved first for a single Cu atom in the O network and showed that a singlet A_{1g} state was formed. The motion of this singlet through the lattice is then described through the t–J model. In contrast to usual heavy fermion materials, we are in the limit here where the singlet binding energy exceeds the Fermi energy of the carriers.

The reduction of the multiband description (1) to the single-band t–J model, is not controlled by a small parameter (e.g. the same physical quantities enter the binding and dispersion of the singlet). This has led to considerable discussion in the literature on the validity and limitations of the reduction from multiband to one-band models (see, for example, Emery & Reiter 1988, 1990; Gooding & Elser 1990; Zhang & Rice 1990*a*, *b*; Shastry 1989). There are interesting questions but of most relevance is the question how good is the reduction in the realistic case when we start with the full hamiltonian (1) and use the parameters in table 1. Recently, Hybertsen *et al.* (1990), have used a cluster method to tackle this problem. They solved the full hamiltonian (1) on a finite cluster of five Cu and 16 O atoms and compared the low energy spectrum with that obtained with the t–J model. They found a good fit between the two spectra both as to the number, degeneracies and energy splittings when they added a t' term, describing next nearest neighbour hopping of the singlet. Their parameters are

$$t = 0.44\ \text{eV}; \quad t' = -0.06\ \text{eV}; \quad J = 0.13\ \text{eV}. \tag{5}$$

Their values have the ratio $J/t < 1$ but not arbitrarily small and the ratio t'/t is also finite.

3. Nuclear resonance studies of the electronic structure

Nuclear resonance techniques are powerful tools to study the low energy electronic structure. In particular they offer site specific probes of the spin densities and through the hyperfine coupling constants of the nature of the electronic wave functions. This is especially true in the case of the cuprate superconductors which have large unit cells with different atomic sites. On the other hand we have discussed above the case that a simplified one-band model is the operative model at low temperatures. In this model the only spin degrees of freedom are Cu^{2+}-moments so that this model can be very directly tested by nuclear resonance methods.

Let us consider the case of the $YBa_2Cu_3O_{7-x}$ compound series since that is the best studied. In particular detailed studies have been made on the ^{63}Cu, ^{17}O and ^{89}Y nuclei and the results have been summarized by Walstedt & Warren (1990). The key assumption made above was that Cu^{2+} ions behave as local $S = \frac{1}{2}$ moments. From this it follows at once that the onsite hyperfine hamiltonian coupling the nuclear and electronic spins has the form

$$H^{(1)}_{\text{hyp}} = \sum_i {}^{63}\boldsymbol{I}_i \cdot A \cdot \boldsymbol{S}_i, \tag{6}$$

where the A coefficients are to be estimated by a comparison to various Cu salts. Since the electronic spin resides in a $3d_{x^2-y^2}$ orbital there is no contact term and the A term is composed of a series of contributions from core polarization, dipole interactions and spin-orbit terms. The resulting hyperfine interactions are anisotropic because of the planar structure. Mila & Rice (1989) estimated the following values for $A_{\parallel}$ (parallel to the c-axis) and $A_{\perp}$ (in the ab-phase)

$$A_{\parallel} \approx 230\ \text{KOe}/\mu_B; \quad A_{\perp} \approx 5\ \text{KOe}/\mu_B. \tag{7}$$

It is impossible to fit the data with such a form and an examination of the structure of CuO_2 planes shows that the well-known phenomenon of transferred hyperfine interaction should be present here. Mila & Rice (1989) did a chemical analysis of the transferred hyperfine interaction and found the leading contribution came from the coupling of a Cu^{2+} electronic spin through σ-overlaps with the O neighbour to a neighbouring 4s-state which in turn has a strong onsite contact interaction with the Cu nuclear spin at that site. In this way a transferred hyperfine interaction occurs between the localized Cu electronic spin on a site and the nuclear spin on a

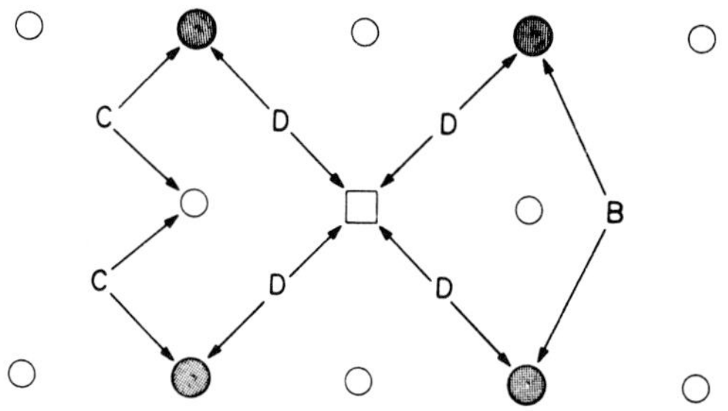

Figure 2. The transferred hyperfine couplings that occur through the 2p-orbital between the Cu^{2+} electronic spins and the ^{63}Cu (B), ^{17}O (C) and ^{89}Y (D) nuclei. ○, O; ●, Cu; □, Y.

Table 2. *Hyperfine parameters for* ^{63}Cu *nuclei*

(All constants in units KOe/μ_B. Hyperfine parameters for ^{63}Cu (2) nuclei in $YBa_2O_3O_7$. The first column gives the values estimated using a chemical analysis (Mila & Rice 1989). In the second, third and fourth columns recent experimental values are quoted.)

	Mila–Rice (1989)	Barrett *et al.* (1990)	Walstedt–Warren (1990)	Imai (1990)
$A_\perp$	5	19	−11	37 ± 5
$A_\parallel$	−228	−214	−188	-158 ± 2
B	40	38	45	39 ± 1
$A_\perp+4B$	165	171	167	195 ± 9
$A_\parallel+4B$	−68	−62	−10	0 ± 4

neighbouring site as illustrated in figure 2. The resultant hyperfine hamiltonian has the form

$$H^{(2)}_{\text{hyp}} = B\sum_i\sum_\tau {}^{63}\boldsymbol{I}_i\cdot\boldsymbol{S}_{i+\tau}, \tag{8}$$

where the sum runs over NN Cu-sites. Mila & Rice (1989) made an estimate of the size expected for the coefficient B and obtained a value $B\approx 40\ KOe/\mu_B$. Since it derives from a contact interaction, it is isotropic in spin space.

This form of the hyperfine hamiltonian can be directly tested. First of all through the Knight shifts $^{63}K_\mu$ which take the form ($\hbar=1$)

$$^{63}K_\parallel = (A_\parallel+4B)\,\chi_0/2\mu_B\,{}^{63}\gamma_n;\ {}^{63}K_\perp = (A_\perp+4B)\,\chi_0/2\mu_B\,{}^{63}\gamma_n, \tag{9}$$

where $^{63}\gamma_n$ is the ^{63}Cu nuclear moment, μ_B the Bohr magneton and χ_0 the spin susceptibility. Many workers (see Walstedt & Warren 1990) have reported that the Knight shift $\parallel c$-axis is very small, essentially zero. This finds a natural explanation in this approach since the combination $(A_\parallel+4B)$ that enters involves an almost complete cancellation between the two terms. For field orientation in the basal plane, it is necessary to have an estimate of χ_0, the spin susceptibility (assumed isotropic in view of the small spin-orbit couplings in Cu ions), to obtain values of the hyperfine constants from the experimental Knight shifts. This leads to some uncertainty in the experimental values. In table 2 we summarize the estimated values, together with those obtained by a detailed analysis of experiments by various groups.

In view of the uncertainties in the parameters the agreement between the theoretical estimates and the experimental values is very satisfactory. Since these hyperfine coupling constants are sensitive to the detailed electronic structure, it is clear that this description which starts from the ionic point of view and treats the hybridization effects perturbatively must be quite good. If there had been substantial changes in the nature of the electronic states then we would expect to see this reflected sensitively in the hyperfine coupling constants.

A second test concerns the doping dependence of the hyperfine constants. Since the $x = 0$ compounds have a substantial hole concentration the question arises what happens when we pass through $x \approx 0.4$ (the 60 K superconductors) to the $x = 1$ case which is an ordered magnetic insulator without hole doping. Again we can ask if there are substantial shifts. The first case that is easy to investigate is the cancellation in the c-axis Knight shift. Since at $x \approx 0.4$ $\chi_0(T)$ develops a strong temperature dependence in the normal state (unlike the case of $x = 0$ which is temperature independent), any change in this cancellation will be easily observed. In fact $K_{\parallel}(T)$ is essentially temperature independent showing that this sensitive cancellation continues in the $x \approx 0.4$ state (see, for example, Walstedt & Warren 1990).

Another independent test of the hyperfine interactions comes from the observation of the antiferromagnetic zero field nuclear resonance. Yasuoka *et al.* (1988) observed this resonance in the $x = 1$ insulator and from their frequency they determine the combination

$$\mu_{\text{eff}} |A_{\perp} - 4B| = 80k \text{ Oe}, \tag{10}$$

where μ_{eff} is the AF ordered magnetic moment. If we take a value $\mu_{\text{eff}} \approx 0.6\,\mu_B$ in agreement with experiment and with theoretical estimates of the ordered moment for $S = \frac{1}{2}$ Heisenberg models in two dimensions, then this yields a value of $|A_{\perp} - 4B| \approx$ 130 KOe/μ_B which we can see from table 1 is in good agreement with the estimates made from the high T_c conductor with $x = 0$. So we come to the conclusion that there is a high degree of consistency in the Cu hyperfine parameters obtained in conducting samples even with different hole concentrations and in insulating samples. Further the values are quite in the expected range for a model of localized Cu^{2+} spins primarily in $3d_{x^2-y^2}$ orbitals. These results speak strongly against a collapse of the local correlation gap upon doping or a collapse of the charge transfer gap either.

This one-band model can be more severely tested by looking at the hyperfine couplings and nuclear resonance on other nuclei. Particularly the ^{17}O nuclei and the ^{89}Y nuclei are good probes and the latter have been studied extensively by Alloul and coworkers (1989). The ^{89}Y nuclei are clearly coupled to the O orbitals through a transferred hyperfine interaction rather than directly to the Cu orbitals which are too far away to overlap substantially. The key assumption in the analysis presented earlier is that the extra holes which are introduced upon doping lie in the σ-bonding O orbitals and are strongly hybridized into local singlets so that the magnetic response is controlled only by the Cu^{2+} spins. This assumption has a clear consequence for the way the resonance properties on the O and Y nuclei should behave. Their response should arise from a transferred hyperfine coupling to the Cu spins and take the form

$$H^{(3)}_{\text{hyp}} = C \sum_{n,\delta} {}^{17}\boldsymbol{I}_n \cdot \boldsymbol{S}_{n+\delta}, \quad H^{(4)}_{\text{hyp}} = D \sum_{n,\delta} {}^{89}\boldsymbol{I}_n \cdot \boldsymbol{S}_{n+\delta}, \tag{11}$$

where the sum over δ runs over the nearest neighbour Cu sites of an O or Y nucleus (see figure 2). This leads at once to the result that the Knight shift observed on the O or Y nuclei should be directly proportional to the spin susceptibility χ_0,

$$^{17}K = C\chi_0/{}^{17}\gamma_n \mu_B; \quad {}^{89}K = 4D\chi_0/{}^{89}\gamma_n \mu_B. \tag{12}$$

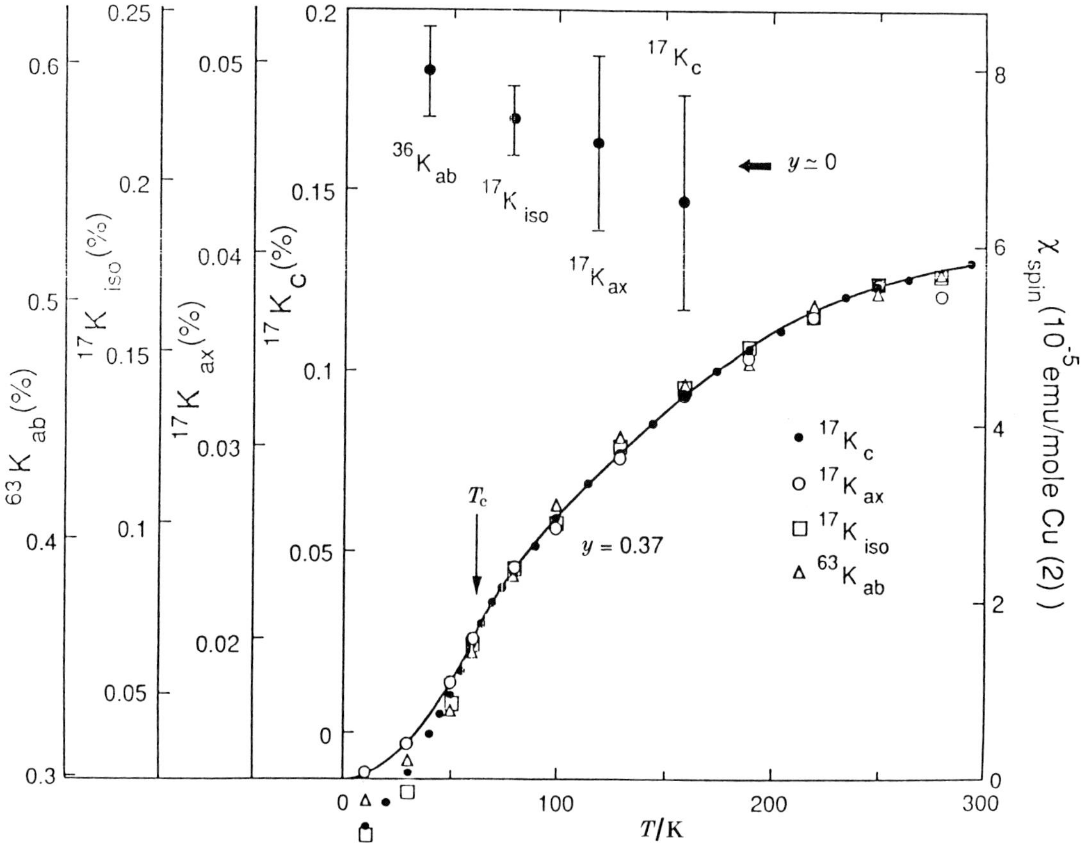

Figure 3. Various components of the Cu and O Knight shifts in $YBa_2Cu_3O_{6.63}$ are plotted against temperature with different vertical scales and origins. The T-independent values of the spin Knight shifts in the $x = 0$ material (i.e. $YBa_2Cu_3O_7$) are also plotted with the same vertical scales. This figure is reproduced from Takigawa *et al.* (1991).

In the case of $x = 0$, since $\chi_0(T)$ is independent of temperature, this is not a very stringent condition. However, for lower doping this condition is much more strict since $\chi_0(T)$ develops a strong temperature dependence even in the normal state. In figure 3 we show the recent results of Takigawa *et al.* (1991). This shows clearly that there is a clear proportionality between the Knight shifts $^{63}K_{\perp}(T)$ on Cu sites, $^{17}K(T)$ on O sites, $^{89}K(T)$ on Y sites and the bulk susceptibility. This is a clear test of the hypothesis that there is a single quantum fluid of strongly hybridized Cu and O orbitals and not two fluids with separate localized Cu spins and itinerant O holes. The latter hypothesis requires a separate contribution particularly on O sites with its own temperature dependence which is in contradiction to the results shown in figure 3. Note, however, that this conclusion is contested by the Grenoble group (Butuad *et al.* 1989; Berthier *et al.* 1990).

The hyperfine hamiltonian (11) for the ^{17}O nuclei describes a purely isotropic Knight shift. In fact a small anisotropy was observed and analysed by Takigawa *et al.* (1989). They showed that it is consistent with a spin density in the σ-bonding 2p-orbitals. This is again consistent with the bonding picture presented above and a further confirmation of the role of the pdσ-hybridization in these compounds.

The relaxation rates of the nuclear resonances are also an important tool to study the low frequency magnetic responses of the quantum fluid. However, we shall not go into these questions and just remark that recently quite detailed analyses of the temperature dependences of these rates have been made (Millis *et al.* 1990; Monien *et al.* 1991*a*, *b*), using a phenomenological form for the wavevector and frequency dependent susceptibility. The very different temperature dependences observed on Cu and on O or Y sites find a natural explanation in terms of predominantly antiferromagnetic fluctuations which are picked up by the Cu nuclear spins but which cancel out at O or Y sites.

4. High energy spectroscopies

There have been many experimental investigations of the cuprate superconductors and we do not attempt a summary of all results. Instead we focus on the question of evidence for the persistence of strong onsite correlations even into the doped materials. Two recent sets of experiments which address this question directly are the electron energy loss spectroscopy (EELS) experiments by Romberg *et al.* (1990) and X-ray absorption spectroscopy (XAS) experiments by Chen *et al.* (1991). Both these experiments in principle measure the same quantity namely the density of states to promote an electron from the filled 1s O core state to an empty 2p-state. In the insulating compounds the empty 2p-states arise due to hybridization of the Cu^{2+} ions with the O neighbours and a low energy peak is found which can be ascribed to this process.

The most interesting results then occur upon hole doping since then one introduces the formal Cu^{3+} sites which are the spin singlets. A new process can now occur in which the excited electron is injected into this state to form a formal Cu^{2+} state. This new process should occur in proportion to the hole doping and an important test of the strong correlation limit, is the question of whether this appears as a separate peak at low energies or merely a broadening of the existing peak. In fact both groups report similar results which show a separate peak growing with hole doping with an energy separation *ca.* 2 eV. A detailed analysis has been made by Hybertsen *et al.* (1990), using the multiband hamiltonian (1) with the parameter values in table 1. They calculate the spectra using a cluster approximation and obtain excellent agreement for the splitting and intensity of the peaks. Since as one discussed earlier, the hamiltonian (1) reduces at low energy to a strongly correlated single-band model, these results are further confirmation of the applicability of the strongly correlated one-band model to the cuprate superconductors. Similar conclusions have been reached by Eskes & Sawatzky (1990) and by Stephan & Horsch (1990).

5. Conclusions

The structure of the CuO_2 planes with four-fold square coordinated Cu ions and two-fold coordinated O ions points directly towards the pdσ-hybridization as the source of the stabilization energy of the planar structure. In this brief review we started from the ionic limit and discussed how by treating this hybridization as the strongest perturbation term we could arrive at a one-band strongly correlated model at low energies. This model in turn makes a series of predictions which are directly tested by nuclear resonance techniques. Also recent high-energy spectroscopy experiments support this model.

The question of the relative weight of the ionic and covalent (or hybridization) description is an interesting and subtle one. Recently Eskes & Sawatzky (1990) examined this question in some detail and attempted a comprehensive look at all experiments which bear on the problem. It is clear that there is strong hybridization between the Cu and O orbitals in these materials (the large values of the Heisenberg coupling J alone attest to that). An intermediate description is probably the most appropriate but this will not negate the single-band model.

In spite of a tremendous amount of activity, it is still not clear whether the simplest single-band t–J model will suffice to describe the superconductivity. The most elegant solution would be to have the high-temperature superconductivity arise as a property of the fixed point of the strongly correlated quantum liquid but at present we know too little to make a definitive conclusion on this point. Empirically the superconducting transition temperature T_c seems to be a function primarily of the hole doping concentration and the interlayer coupling.

However, in a conference such as this devoted to the interplay between electronic properties and crystal structure, it is appropriate to discuss some recent experiments which point to a curious but clear influence of a subtle change of structure. Axe *et al.* (1989) and Maeno *et al.* (1990) have found that in the $La_{2-x}Ba_xCuO_4$ series there is a region of a stability of second tetragonal phase which has a dramatic effect to reduce the value of T_c. Yet this phase is only very slightly different to the orthorhombic phase. The difference occurs through the different pattern of O-octahedra tilting in the two phases. How can such a subtle change in crystal structure influence T_c dramatically? This is another puzzle to add to the list as we try to unravel the way from the crystalline structure to the bonding pattern and then to the electronic structure and finally the superconductivity of the cuprate materials.

We are grateful to very many colleagues for informative discussions on these topics, especially, P. W. Anderson, M. S. Hybertsen, M. Schlüter, H. Monien, D. Pines, P. C. Hammel, M. Takigawa, R. E. Walstedt and W. W. Warren.

References

Alloul, H., Ohno, T. & Mendels, P. 1989 ^{89}Y NMR evidence for a Fermi liquid behaviour in $YBa_2Cu_3O_{6+x}$. *Phys. Rev. Lett.* **63**, 1700–1703.

Anderson, P. W. 1987 The resonating valence bond state in La_2CuO_4 and superconductivity. *Science, Wash.* **235**, 1196–1198.

Anderson, P. W. 1988 50 years of the Mott phenomenon: insulators, magnets, solids and superconductors as aspects of strong repulsion theory. In *Frontiers and borderlines in many-particle physics* (ed. R. A. Broglia & J. R. Schrieffer). Amsterdam: North Holland.

Axe, J. D., Moudden, A. H., Hohlwein, D., Cox, D. E., Mohanty, K. M., Moodenbaugh, A. R. & Xu, Y. 1989 Structural phase transformations and superconductivity in $La_{2-x}Ba_xCuO_4$. *Phys. Rev. Lett.* **62**, 2751–2754.

Barrett, S. E., Durandt, D. J., Pennington, C. H., Slichter, C. P., Friedman, T. A., Rice, J. P. & Ginsberg, D. M. 1990 ^{63}Cu Knight shifts in the superconducting state of $YBa_2Cu_3O_7$. *Phys. Rev.* B **41**, 6283–6287.

Bedell, K. S., Coffey, D., Meltzner, D. E., Pines, D. & Schrieffer, J. R. 1989 High temperature superconductivity. *Proc. Los Alamos Symp.* New York: Addison-Wesley.

Bednorz, J. G. & Müller, K. A. 1986 Possible high T_c superconductivity in the Ba–La–Cu–O system. *Z. Phys.* B **64**, 189–193.

Berthier, C., Berthier, Y., Butuad, P., Horvatic, M. & Segranson, P. 1990 ^{17}O NMR investigation of the electronic structure of high-T_c superconducting oxides. In *Electronic properties of high-T_c superconductors* (ed. H. Kuzmany). Springer-Verlag.

Butuad, P., Horvatic, M., Berthier, Y., Segranson, P., Kitaoka, Y. & Berthier, C. 1990 ^{17}O NMR in $YBa_2CuO_{6.6}$ discrimination between *t–J* and 2 band models. *Physica* C **166**, 301–309.

Chen, C. T., Sette, F., Ma, Y., Hybertsen, M. S., Stechel, E. B., Foulkes, W. M. C., Schlüter, M., Cheong, S. W., Cooper, A. S., Rupp, L. W., Batlogg, B., Soo, Y. L., Ming, Z. H., Krol, A. & Kao, Y. H. 1991 Electronic states in $La_{2-x}Sr_xCuO_4$ probed by soft X-ray absorption. *Phys. Rev.* **66**, 104–107.

Chen, C. X., Schüttler, H. B. & Fedro, H. J. 1990 Hole excitation spectra in cuprate superconductors: a comparative study of single- and multiple-bond strong coupling theories. *Phys. Rev.* B **41**, 2581.

Coffey, D., Rice, T. M. & Zhang, F. C. 1990 (In preparation.).

Emery, V. J. 1987 Theory of high-T_c superconductivity in oxides. *Phys. Rev. Lett.* **58**, 2794.

Emery, V. J. & Reiter, G. 1988 Quasiparticles in the CuO planes of high-T_c superconductors: an exact solution for a ferromagnetic background. *Phys. Rev.* B **38**, 11938–11941.

Emery, V. J. & Reiter, G. 1990 Reply to 'validity of the *t–J* model'. *Phys. Rev.* B **41**.

Eskes, H. & Sawatzky, G. 1988 Tendency towards local spin compensation of holes in high-T_c copper compounds. *Phys. Rev. Lett.* **61**, 1415.

Eskes, H. & Sawatzky, G. 1990 The influence of hybridization on the spin and charge distribution in the high-T_c superconductivity. *Proc. IWEPS '90, Kirchberg, Austria.*

Gooding, R. J. & Elser, V. 1990 Infinite $U_d\,U_p$ ground state of the extended Hubbard model. *Phys. Rev.* B **41**, 2557–2559.

Hybertsen, M. S., Schlüter, M. & Christensen, N. F. 1989 Calculation of Coulomb interaction parameters for La_2CuO_4 using a constrained-density-functional approach. *Phys. Rev.* B **39**, 9028–9034.

Hybertsen, M. S., Stechel, E. B., Schlüter, M. & Jennison, D. R. 1990 Renormalization from density functional theory to strong coupling models for electronic states in Cu–O materials. *Phys. Rev.* B **41**, 11068.

Hybertsen, M. S., Stechel, E. B., Foulkes, W. M. C. & Schlüter, M. 1990 (In the press.)

Imai, T. 1990 Analysis of nuclear relaxation experiments in high-T_c oxides based on the Mila–Rice hamiltonian. ISSP preprint.

Maeno, Y., Kahehi, N., Odagawa, A. & Fujita, T. 1990 Structural transition and superconductivity in $(La_{1-z}Ba_{x-y}Sr_y)_2CuO_4$. *Proc. LT19, Sussex, U.K.*

McMahan, A. K., Martin, R. M. & Satpathy, S. 1988 Calculated effective hamiltonian for La_2CuO_4. *Phys. Rev.* B **38**, 6650–6655.

Mila, F. 1988 Parameters of a Hubbard hamiltonian to describe superconducting Cu oxides. *Phys. Rev.* B **38**, 11358–11361.

Mila, F. & Rice, T. M. 1989 Analysis of magnetic resonance experiments in $YBa_2Cu_3O_7$. *Physica* C **157**, 561–570.

Millis, A. J., Monien, H. & Pines, D. 1990 Phenomenological model of nuclear relaxation in the normal state of $YBa_2Cu_3O_7$. *Phys. Rev.* B **42**, 167–180.

Monien, H., Pines, D. & Takigawa, M. 1991*a* On the application of antiferromagnetic Fermi liquid theory to NMR experiments on $YBa_2Cu_3O_{6.63}$. *Phys. Rev.* B **43**, 258–274.

Monien, H., Monthoux, P. & Pines, D. 1991*b* On the application of antiferromagnetic Fermi liquid theory to NMR experiments in $La_{1.85}Sr_{0.15}CuO_4$. *Phys. Rev.* B **43**, 275–285.

Ramšak, A. & Prelovšek, P. 1989 Comparison of effective models for CuO_2 layers in oxide superconductors. *Phys. Rev.* B **40**, 2239–2243.

Romberg, H., Alexander, M., Nücker, N., Adelmann, P. & Fink, J. 1990 On the insulator–metal transition in $La_{2-x}Sr_xCuO_4$. *Phys. Rev.* B **42**, 8768–8771.

Schmidt, H. J. & Kuramoto, Y. 1990 Four spin interaction as an effective interaction in high-T_c copper oxides. Preprint.

Shastry, B. S. 1989 *t–J* model and nuclear relaxation in high-T_c materials. *Phys. Rev. Lett.* **63**, 1288.

Stephan, W. & Horsch, P. 1990 Calculations of photoemission spectra for the *t–J* model and the extended Hubbard model. In *Dynamics of magnetic fluctuations in high-T_c* superconductors (ed. G. Reiter). New York: Plenum. (In the press.)

Takigawa, M., Hammel, P. C., Heffner, R. H., Fisk, Z., Ott, K. C. & Thompson, J. D. 1989 ^{17}O NMR study of local spin susceptibility in aligned $YBa_2Cu_3O_7$ powders. *Phys. Rev. Lett.* **63**, 1865–1868.

Takigawa, M., Reyes, A. P., Hammel, P. C., Thompson, J. D., Heffner, R. H., Fisk, Z. & Ott, K. C. 1991 Cu and O NMR studies of the magnetic properties of $YBa_2Cu_3O_{6.63}$. *Phys. Rev.* B **43**, 247–257.

Walstedt, R. E. & Warren, W. W. Jr 1990 Nuclear resonance properties of $YBa_2Cu_3O_{6+x}$. *Science, Wash.* **248**, 1082–1087.

Yasuoka, H., Shimizu, T., Ueda, Y. & Kosuge, K. 1988 Observation of antiferromagnetic nuclear resonance of Cu in $YBa_2Cu_3O_6$. *J. Phys. Soc. Japan* **57**, 2659–2662.

Zhang, F. C. & Rice, T. M. 1988 Effective hamiltonian for the superconducting Cu oxides. *Phys. Rev.* B **37**, 3754–3762.

Zhang, F. C. & Rice, T. M. 1990*a* Reply to 'infinite U_d U_p ground state of the extended Hubbard model'. *Phys. Rev.* B **41**, 2560.

Zhang, F. C. & Rice, T. M. 1990*b* Validity of the *t–J* model. *Phys. Rev.* B **41**, 7243.

Discussion

L. J. SHAM (*University of California, San Diego, U.S.A.*). Professor Rice has given a very persuasive account of how a strongly correlated electron model can explain a number of experimental observations in copper oxides, including NMR, etc. Does this also mean that the observations necessarily imply strong correlation? If so, how is one to think of the RPA calculations of Scalapino *et al.* which explain the NMR data or of the LDA calculations of the Fermi surfaces which are still quite satisfactory?

T. M. RICE. Taking the NMR case first, the RPA calculations refer mostly to the strong AF correlations which show up in the relaxation rate. I believe they may be more applicable to the case of large hole doping where there is evidence that a Landau–Fermi liquid forms when $T_c \rightarrow 0$. For smaller hole concentration, e.g. the case of $YBa_2Cu_2O_{6.6}$ which has a strongly temperature dependent, $\chi_0(T)$, there is definitely a need for a strongly correlated model to explain the spin pairing over a large region in temperature that appears in $\chi_0(T)$. RPA gives only a constant value for the uniform susceptibility, $\chi_0(T)$.

Turning to the angle-resolved photoemission experiments, in the crudest sense they measure the value of $n(k)$, the Bloch momentum expectation value when one integrates up to the Fermi energy. We know from the one-dimensional Hubbard model that $n(k)$ drops off sharply at the Luttinger value of k_F and I expect something similar will occur in two dimensions. The detailed spectral form is another matter which requires more theory but the experiments are not obviously consistent with Landau–Fermi liquid theory.

A. O. E. ANIMALU (*University of Nigeria, Nsukka, Nigeria*). How does Professor Rice's model explain the observation that substitution of Cu^{II} by Zn or Ni leads to lowering of T_c, more dramatically for Zn with ($3d^{10}$) configuration than for Ni with ($3d^8$) configuration?

T. M. RICE. It is important to obtain the high-T_c superconductivity to have the Cu O_2 planes as perfect as possible and to dope them as unintrusively as possible. Dopants, such as just mentioned, which substitute on Cu planar sites are strong perturbations of the planes and destroy the integrity of the planes and in this way lower T_c.

L. M. FALICOV (*University of California, Berkeley, U.S.A.*). How can Professor Rice justify a treatment of individual CuO_2 layers when these are strongly charged? The stoichiometry of $Cu^{2+}+2O^{2-}$ yields two electrons per formula unit, a strongly charged layer indeed.

T. M. RICE. Empirically there is evidence for strong anisotropy parallel and perpendicular to the CuO_2 planes. Therefore it seems that the hopping and interactions between planes are quite small and we need to start with a good understanding of a single layer. However, the interplanar interactions may still in the end determine, T_c as for example Professor Anderson has argued on empirical grounds.

Pseudopotentials and the theory of high T_c superconductivity

By P. W. Anderson

Joseph Henry Laboratories of Physics, Jadwin Hall, Princeton University, Princeton, New Jersey 08544, U.S.A.

I present two very different uses of the idea of pseudopotentials in the theory of cuprate superconductors. In the first, the 'chemical pseudopotential' scheme is used to set up the underlying Hubbard model which is appropriate for these substances; in the second, I show that the conventional multiple-scattering technique for constructing an effective scattering length theory does not converge for the two-dimensional Hubbard model.

When working on the subject which has been engrossing me for the past four years, high T_c superconductivity, I often feel that I should be disqualified from the competition because of the fact that I carry a number of concealed weapons on me, not the least of which is the understanding of pseudopotentials which I owe to V. Heine.

Yes, Virginia, contrary to the conventional wisdom, there is a theory of high T_c superconductivity which gives a highly satisfactory explanation of most of the facts which are facts, though it is not exactly what you have been hearing. But it is as long, deep and devious as a physical theory can be, and I here relate a couple of parts of it which are related to the theory of pseudopotentials, which I learned from Volker and his friends and collaborators.

There are two places where pseudopotentials play an all-important role in the theory of high T_c: in the understanding of the parameters of the Hubbard model which underlies the whole phenomenon; and in the understanding of the divergence which, to our surprise and delight, invalidates the use of Fermi liquid theory to describe the metallic state in this model.

I will not bore you with the well-known structures of these materials, which in any case you have already seen. The operative elements are the CuO_2 planes (see figure 1), which can occur with or without the apical oxygens which complete an irregular pyramid or octahedron of O about each Cu. Aside from being good indicators of the electronic structure on the Cu, these are not electronically relevant.

All of the interesting action, in fact, takes place in the rather strongly bonding σ-bands of the CuO_2 structure. The π bands are entirely full at the physical occupancy, and do not contribute to bonding; so one is left with d_z^2 and $d_{x^2-y^2}$ orbitals on the Cu – with a slight but NMR-relevant admixture of Cu4s – and p_σ on the oxygen.

The 'chemical pseudopotential' method which Dave Bullett and I originated some years ago is the simplest way of setting up an effective hamiltonian for the relevant bands. If you remember, the idea was to rely on the chemical observation that in compounds like this with relatively large interband gaps, LCAO theory gives a good qualitative account, and what the chemists call 'extended Hückel' even a quantitative one, of the electronic structure. We search for the best atomic-like orbitals by using non-orthogonal orbitals and pseudizing away the biggest part of the effects of the neighbouring atoms. We have two systems of equations, a pseudopotential equation for the atomic orbitals and a derived effective 'Hückel' type hamiltonian for the actual energy levels.

In the present case, there's a detail which must be taken care of: we must somehow make a hole–particle transformation because we are interested only in the top of the band: in how extra holes repel each other rather than extra electrons. For holes, atom cores represent strong repulsive hard cores, but as far as I can see pseudization can be used to mitigate the effects of hard cores just as well; this was in fact the original use of pseudopotentials.

If we look at the band upside down, in this way, we realize that it is not at all interesting to include the bonding orbitals, which are simply part of a lot of high-energy junk taking maximum advantage of the (now repulsive) ion cores. There is no way that they could, in some esoteric fashion, contribute to an effective attraction between holes: all of the nonsense about 'extended Hubbard models' just drops away. What is left is to solve for the antibonding pseudo-Wannier functions of $Cu_{x^2-y^2}$ and Cu_{z^2} symmetry, and to study the shift in their energies as we vary the occupancy from zero to one to two holes. Dave Bullett has developed a kind of two-stage version of this procedure: first we set up non-orthogonal atomic orbitals which give the parameters of an effective hamiltonian by

$$(T+V_{\mathrm{Cu}})\varphi_{\mathrm{Cu}}-\sum_{\mathrm{O}}[(\varphi_{\mathrm{O}}|V_{\mathrm{O}}|\varphi_{\mathrm{Cu}})\varphi_{\mathrm{O}}-V_{\mathrm{O}}\varphi_{\mathrm{Cu}}]=E_{\mathrm{Cu}}\varphi_{\mathrm{Cu}},$$

$$(T+V_{\mathrm{O}})\varphi_{\mathrm{O}}-\sum_{\mathrm{Cu}}[\varphi_{\mathrm{Cu}}|V_{\mathrm{Cu}}|\varphi_{\mathrm{O}})\varphi_{\mathrm{Cu}}-V_{\mathrm{Cu}}\varphi_{\mathrm{O}}]=E_{\mathrm{O}}\varphi_{\mathrm{O}},$$

and $$\mathscr{H}_{\mathrm{O-Cu}}=\int\varphi_{\mathrm{Cu}}V_{\mathrm{Cu}}\varphi_{\mathrm{O}}\,\mathrm{d}^3r,\quad \mathscr{H}_{\mathrm{Cu-O}}=\int\varphi_{\mathrm{O}}V_{\mathrm{O}}\varphi_{\mathrm{Cu}}\,\mathrm{d}^3r.$$

(Note the simplifying difference in the way the bonding matrix elements are calculated.)

Then we set up, using these parameters, a pseudo-Wannier bonding or anti-bonding function which, again, need not be orthogonalized, and rather approximately (but the corrections can be calculated simply) is an eigenfunction of the 'cluster' hamiltonian of a Cu and its neighbour oxygens:

$$\Psi^{\mathrm{PW}}_{\mathrm{hole}}=\alpha\varphi^{x^2-y^2}_{\mathrm{Cu}}-\beta\sum_{1}^{4}\varphi^{\mathrm{p}^{\sigma}}_{\mathrm{O}i}(-1)^i.$$

I obviously neither have, nor am going to, explicitly do calculations for this system: much more able people, like Michael Schluter or Dave Bullett, have done them for me. But it is very interesting to think about where the parameter 'U' comes from. Although the above is a one-electron calculation, what we can do is to repeat the local one-electron calculation for two different local occupancies.

This then becomes virtually identical to the kind of local cluster theory which George Sawatsky or Jim Allen use to understand PES results or to Gunnarson's LCI theory. If we, for instance, are calculating for the first hole to be removed from the cluster, the Cu d level is high (attractive for holes), about 3–4 eV above the O_{2p} level.

The net result is an energy level which is dominantly Cu $d_{x^2-y^2}$, not strongly antibonding but somewhat (since V_{pd} is only about comparable with the p–d splitting).

Removing one of these electrons, the next electron sees a Cu d level lower than the oxygen (but not much lower, since the electron came *ca.* 30% from oxygen) and we get an antibonding level which is somewhat more O than Cu; the difference is, effectively, 'U', which is then never bigger than the smaller Cu–O splitting, since the electron can at worst simply shift from Cu to O. But it is important to realize that these are not orthogonal d and p orbitals, and that even a pure d or p level still has some bonding character: the Jahn–Teller bond still has considerable strength even in the nominally purely ionic system. Thus the shift in atomic provenance does not mean that the electrons are from different bands, as is often mistakenly assumed.

A much more interesting and, as yet, controversial use of the pseudopotential concept and of all the experience with scattering wave functions and phase shifts which we garnered in the 1960s, is in the central proof of the non-Fermi liquid character of the two-dimensional Hubbard model. In fact, I would stick my neck out and say that no single-band interacting Fermi system in two dimensions is a Landau–Fermi liquid.

One of the most disturbing things about this pesky subject is the way it has of destroying prejudices which we all seem to have unconsciously enshrined as facts, over the years. I cribbed from Joseph Ford via James Gleick's book a marvellous quotation from Tolstoy on this subject:

> I know that most men ... can seldom accept even the ... most obvious truths, if they be such as would oblige them to admit the falsity of conclusions which they have delighted in explaining to colleagues, which they have proudly taught to others, and which they have woven ... into the fabric of their lives.

Just such a prejudice is that the Landau theory is derived from many-body perturbation theory, which is fundamental and exact.

In a sense, Landau theory can be the more fundamental, since it can be derived from a consistent renormalization group procedure, while the perturbation theory is incomplete because, as normally carried out, it does not contain within itself the proper determination of the particle–particle scattering vertex which it uses in the Hartree terms which begin the series. I will save for later the first of these thoughts, and only talk here about the determination of the pseudopotential for particle–particle interaction.

This is, of course, a very old problem extending back into the history of many-body theory. Brueckner got his name on the idea that hard core interactions had to be renormalized away by replacing the simple interaction $V_{kk'}$ by an effective interaction for which he proposed the use of the scattering matrix $T_{kk'} = V_{kk'} + (VG_0V)_{kk'} + \ldots$ Lee, Yang and Huang gave a fairly rigorous treatment of this process in the low-density limit, and seemed to convincingly show that an 'effective scattering length' theory was the appropriate solution. The meaning of this resummation is the observation that in the actual region where two particles are close to each other, unless the interactions are very long-range indeed it is never valid to assume that the wave-functions are unperturbed, and so one must in some sense solve a local Schrödinger equation for the scattering states, to get the correct asymptotic behaviour of the wave-functions. In a sense, every pair of particles experiences a Jastrow-like 'hole' in the near region. Where everything behaves

Figure 1

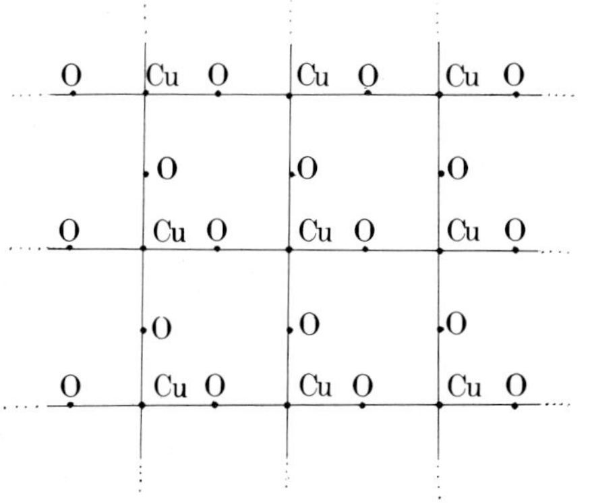

Figure 2

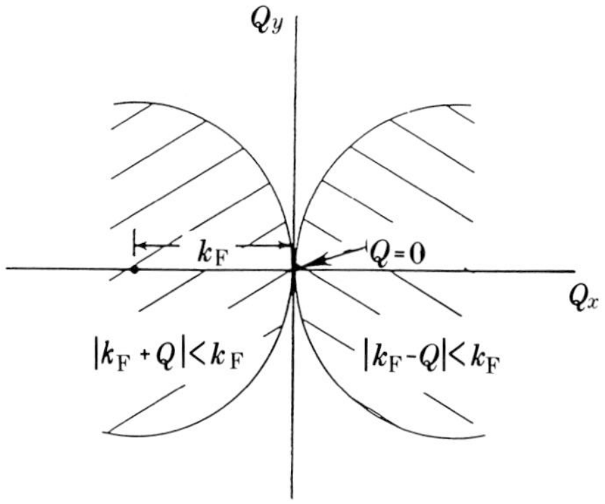

Figure 1. CuO_2 lattice.
Figure 2. Region of recoil momentum Q excluded by the Pauli principle.

regularly, as in three dimensions and for moderate interactions and free-electron bands, for two particles with relative momentum Q

$$T = (e^{i\eta} \sin \eta)/Q \sim \eta/Q$$

for small Q and phase-shift. The phase shift η is Qa, where a is the scattering length, so V is replaced by a finite, simple pseudopotential which merely expresses the fact that there is an additional kinetic energy caused by the small restriction on Hilbert space due to the atom cores. If this is the case, everything remains finite and Fermi liquid theory works. In particular, any one particle k does not cause an appreciable scattering phase shift for any other particle k', and it is consistent to treat the quasiparticles as though they occupied exactly the same volume in Hilbert space as the corresponding free electron gas. (The actual shift is *ca.* a/L where L is the sample size.) It is important, however, to realize that when we solve the scattering Schrödinger equation to get the local wave function, we are always doing so in the presence of boundary conditions: we are no longer dealing just with free plane waves but either with real standing waves $\sin(kr)$, $\cos(kr)$, or with ingoing and outgoing partial waves. Thus the naive picture of a Hartree term as just a $q = 0$ scattering is now false, and it includes both forward and backward scattering.

Not long thereafter Bloom took a look at the same problem in one and two dimensions, and discovered that the situation was by no means as simple in the low density limit (which is, I repeat, the only limit where rigorous results exist). In fact, in this limit the scattering length diverges, in one dimension like the sample size L and in two dimensions like $L/\ln L$, so that in the limit $Q \to 0$, $\eta_{1D} = \pi$ and $\eta_{2D} \sim 1/\ln L$. This means that the wave-function correction extends throughout the sample in one dimension since, after all, any scattering at all effectively changes the boundary conditions in one dimension and halves the effective length of the sample. Correspondingly, there are no Fermi liquids in one dimension, even for weak repulsive forces.

In two dimensions, the problem of the T-matrix becomes quite tricky. The phase shift, however, tends to 0 as $1/\ln L$, and Galitskii showed that in that case a Fermi liquid theory can indeed be made to – barely – converge.

These problems in one and two dimensions have to do with, not the short-range part of the solution of the Schrödinger equation, but the long-range part: the behaviour 'on the energy shell' which was always a thorn in the side of the multiple-scattering theory. In solving the T-matrix equation $T = V + VG_0 T$ or $T = (1-VG_0)^{-1} V$, one has to worry in detail about what happens at the pole in G_0 corresponding to the state actually being considered, or, specifically, what boundary conditions do you apply? (Again, boundary conditions!) Another way of seeing the problem is to look at the low-momentum solutions to scattering theory, and realize that these are simply $(x-a)$ in one dimension and $\ln r/a$ in two, both of which have relevant long-range as well as short-range behaviour.

What I discovered, and what actually is really terribly obvious, is that in two dimensions, for a cut-off band or in the Hubbard model, the Galitskii calculation becomes even more singular when one takes the exclusion principle into account. In effect, the phase-shift for zero density vanishes because the particles can recoil with any low energy: but if there is a filled Fermi sea, these low-energy recoils are excluded, the long-range part of the $\ln r/a$ wave-function is cut off, and the result is a finite phase shift, caused by the resulting finite logarithmic derivative of the relative wave-function.

The way of showing this is very easy, if one recognizes the vital role of boundary conditions and the energy shell. The relevant calculation is that of the actual energy shift of two particles with zero relative momentum, presuming that the two particles are near the Fermi surface, i.e. in the $K_{\text{tot}} = 2k_F$ singlet particle–particle channel. If we wish to calculate the energy shift, we must, of course, occupy the two relevant states, but we cannot allow them to recoil into states already occupied by other electrons, because of the exclusion principle. The equation which we must satisfy, then, is if we let

$$E_Q = \epsilon_{k+Q} + \epsilon_{k-Q}$$

and

$$\Psi_{\text{scatt}} = \sum_Q a_Q c^+_{k+Q\uparrow} c^+_{k-Q\downarrow} |0\rangle,$$

is

$$\mathcal{H}\Psi_{\text{scatt}} = E\Psi_{\text{scatt}} = \frac{U}{L^2} \sum_{QQ'} a_Q c^+_{k+Q'\uparrow} c^+_{k-Q'\downarrow} + \sum_Q E_Q a_Q c^+_{k+Q\uparrow} c^+_{k-Q\downarrow},$$

or

$$(E-E_Q) a_Q = \frac{U}{L^2} \sum_Q a_{Q'}, \quad 1 = \frac{U}{L^2} \sum_{Q'} \frac{1}{E-E_{Q'}}.$$

If there is no exclusion principle, life is simple. In two dimensions, near the bottom of the band, states are evenly spaced,

$$E_Q = E_0 + Q^2, \quad (Q = n_x \pi/L, n_y \pi/L).$$

The lowest eigenvalue lies near E_0 so we have

$$\frac{1}{E-E_0} = \frac{L^2}{U} + \sum_{Q=\pi/L}^{Q_{\max}} \frac{1}{Q^2-E} = \frac{L^2}{U} + \frac{L^2}{\pi^2} \ln \frac{L}{a}$$

and, as Bloom found, $(E-E_0)/(E_1-E_0) \approx 1/(\ln La^{-1})$. Thus asymptotically, the energy in the relative motion must be slightly higher, and in the scattered wave function there is a small shift in Q, but this vanishes as $L \rightarrow \infty$.

Now consider the case where there is a finite density, $k_F > 0$. Now, if Fermi liquid theory were correct, the states $k+Q$ will be occupied with density $n_F(\epsilon_{k+Q})$ which (see figure 2) will exclude recoils in all but a narrow band of final states of density $Q^2/\pi k_F^2$. In essence, the wave function of relative motion $\ln r/a$ contains, in its longer-range parts, momenta which are not available because already occupied. Fortunately, because of the exclusion principle the sum no longer depends in detail on the low-energy end so that the treatment of Qs near zero is unimportant; and we simply remark that the exclusion principle cuts off the logarithm at k_F, not π/L:

$$\frac{\delta E}{\text{spacing}} = \frac{E-E_0}{E_1-E_0} \sim \frac{1}{\ln(k_F a)} \sim \eta_{l=0}.$$

The phase-shift as $Q \rightarrow 0$ is finite! Thus the same difficulties which prevent the validity of FLT in one dimension are completely valid here.

It is not appropriate here to go into too great detail as to what this does to the ground-state wave function.

What we can be sure of is that the two-dimensional Hubbard model has no normal state which is a Fermi liquid, even for low density or weak coupling and *a fortiori* for strong coupling and high density as in the CuO_2 planes. How this fact leads to the rest of the phenomenology of high T_c superconductivity is pretty well understood, but lengthy; and this meeting is not about that. I do believe that this is a generalizable statement which applies as well to SDW and CDW-forming systems like $NbSe_2$ and TaS_2, surface bands on semiconductors, or quasi-two-dimensional organics: we will just have to face the fact that we must rethink the physics of all of these restricted dimensionality systems, sometime quite soon. I hardly need to tell Volker, of course, that the conventional 'nesting Fermi surface' ideas never really carried plausibility to properly observant theorists.

With these disturbing but exciting thoughts I shall close. TCM, as a subject, is alive and not just well but kicking; to mix metaphors, the cat has not only not got all the cream, we seem to have tapped into a wholly new and unexpected set of cows.

Discussion

J. N. MURRELL (*University of Sussex, U.K.*). Is there a specific chemical feature of a single CuO_2 two-dimensional sheet that would lead to long-range electron pair correlation for HTSC structures? In my view the fact that these sheets are, in chemical terms, alternant systems systems (whose hamiltonian matrices can be blocked in the form

$$\begin{bmatrix} 0 & \boldsymbol{h} \\ \boldsymbol{h} & 0 \end{bmatrix}$$

may have significance, for the lowest eigenfunctions of such systems appear to have the electron correlation needed.

P. W. ANDERSON. (1) The bands do not have very good alternant symmetry because of O–O overlap effects. (2) In my opinion, supported by heuristics and by observations on one-layer Bi compounds, single layers are very interesting but not true superconductors. Elsewhere I have suggested that they are '$T_c = 0$ superconductors'. (3) The interesting region for superconductivity is large doping when the commensurability question ('nesting', 'spin gaps', etc., and ideas involving AB bonding between two sublattices) is no longer relevant.

T. M. RICE (*Zurich, Switzerland*). In a recent *Physical Review Letter*, Engelbrecht & Randeria examined the conventional perturbation theory of a low density repulsive Hubbard model and found that forward scattering did not invalidate Landau–Fermi liquid. How does this relate to Professor Anderson's work?

P. W. ANDERSON. We have been examining this paper, especially A. Georges and myself. The key point is that the problem occurs before perturbation theory, and cannot be found by resummation of the conventional diagrams assuming the vertex is not divergent, at least if the diagrams are done in the conventional way. The result can probably be demonstrated by calculating the Hartree self-energy shift accurately taking into account the detailed structure of Γ. In the conventional theory this term is calculated by a coupling constant integration rather than directly, but we now believe the theory is non-analytic in 'g' at every point. This integration is used, among other things, to avoid treating the behaviour of T on the energy shell (at the pole) in detail, which is what is necessary. Our view is that Randeria–Engelbrecht is simply irrelevant to the point.

The conventional theory, to put it another way, contains an infinite renormalization of the vertex due to recoil which is normally unexamined. Section 5 of AGD shows that each order of perturbation theory contains a correction to the assumed pseudo-potential a, the sum of which is a bit we have examined and found divergent.

V. HEINE (*Cambridge, U.K.*). Could Professor Anderson please comment on Hugenholtz's old problem, namely that perturbation theory for the electron gas cannot possibly converge in one, two or three dimensions. The argument was that a power series has a circle of convergence, i.e. the same range of convergence for positive and negative interactions. Now an attractive Coulomb potential cannot give a convergent perturbation series and therefore the actual repulsive electron–electron interaction cannot give a convergent series either.

P. W. ANDERSON. This is a real problem for Fermi liquid theory, discussed at length in AGD, §12. Fortunately, for repulsive forces the Cooper singularity renormalizes the $k, -k$ scattering to zero; this renormalization is non-analytic but *improves* convergence. It does lie outside naive perturbation theory.

Density functionals beyond the local approximation

By L. J. Sham

Department of Physics, University of California, San Diego, La Jolla, California 92093-0319, U.S.A.

A review of the current effort in improvement over the local density approximation is given. Within the density functional theory, the exchange-correlation energy and potential may be unambiguously defined. Based on the field theoretical expressions for them, approximations for classes of systems and approximate evaluations for specific solids are critically reviewed. Further lines of development are discussed. Relation to the quasi-particle energies is explored.

1. Introduction

The density functional theory (Hohenberg & Kohn 1964; Kohn & Sham 1965) in the local density approximation gives a simple way of utilizing the knowledge of the exchange-correlation effects of the electron gas in real atoms, molecules and solids. Through the effort of many workers in this field, the theory has been thoroughly tested in a wide range of systems including atoms, molecules, solids and nuclei and found particularly useful for ground-state properties (Schlüter & Sham 1982; Lundqvist & March 1983; Parr & Yang 1989). There is of course room for improvement. There are two separate aspects. One is the improvement of the energy functionals for the ground-state properties. The other is how the energies of low-lying excited states may be found within the context of the density functional theory.

In §2 the definitions of the exchange and correlation energy functionals and potentials are given in the context of the density functional theory and the motivation of the particular definitions is shown to be the transformation of a many-particle problem to an effective one-particle one. This does not eliminate the interacting particle physics but only moves it to the construction of the effective potential. Section 3 reviews a formal expression for the exchange-correlation energy. Section 4 explains the difference between the density functional exchange and the Fock exchange in the Hartree–Fock approximation. Section 5 explores a number of approaches that go beyond the local density approximation (LDA).

The low-lying excited state energies may be viewed as quasi-particle energies. The density functional equation for constructing the ground-state density naturally has a set of single-particle energies. Whether they can represent the quasi-particle energies has been investigated. In general the two sets are not related. None the less, the density functional theory is shown to be a useful framework to construct the self-energy. The insulator or semiconductor band gap is a special case in which it is related to the ground-state energies of the system and of the system with the addition or removal of an electron. The discontinuity of the exchange-correlation potential on adding an electron in an insulator plays a prominent role in the determination of the band gap (§6). The Fermi surface is formally not given by the density functional eigenvalues but is in practice remarkably well approximated by the LDA in a disparate range of metals (§7). The reason remains to be explored. The final section summarizes the situation that improvement over the LDA involves considerations of specific systems with the loss of the simplicity and universality of the LDA.

2. Definition of the exchange-correlation energy functional

The density functional theory has three key features: (1) any property of a many-electron system is a functional of the ground state density; (2) for a given external potential, the ground state energy is a variational minimum at the correct density; (3) the density is determined by a one-particle Schrödinger equation with an effective potential.

In the consideration of the class of all many-electron systems, which share the common attributes of the electron mass, Fermi statistics, and the same Coulomb interaction between a pair of electrons, the position-dependent external potential $v(r)$, i.e. the potential experienced by each electron due to the fixed nuclei, characterizes a particular system. A property of the system is said to be a functional of $v(r)$. In particular, the density $n(r)$ is a functional of $v(r)$, it being understood that either the total number of electrons or the chemical potential is given. The theorem of Hohenberg & Kohn (1964) asserts that the external potential is a functional of the ground-state density, apart from a trivial constant.

It follows that the ground-state energy is a functional of the density. It may be separated into two terms:

$$E = T_s[n] + U[n], \tag{2.1}$$

where $T_s[n]$ is the kinetic energy functional of a non-interacting electron system with the same density distribution. By comparing the variational equation

$$\delta T_s/\delta n + \delta U/\delta n = \mu, \tag{2.2}$$

μ being the chemical potential, with the corresponding one for the non-interacting system, one arrives at the one-electron Schrödinger equation

$$[-\tfrac{1}{2}\nabla^2 + v_{\text{eff}}(r)]\,\psi_j(r) = \epsilon_j\,\psi_j(r), \tag{2.3}$$

which determines the density

$$n(r) = \sum_j \theta(\mu - \epsilon_j)\,|\psi_j(r)|^2. \tag{2.4}$$

The effective one-electron potential is given by

$$v_{\text{eff}}(r) = \delta U/\delta n. \tag{2.5}$$

Formally, the density functional theory has reduced the many-body problem to the solution of a one-particle Schrödinger equation. The many-body problem now consists in the construction of the energy functional $U[n]$ or its functional derivative $v_{\text{eff}}(r)$. The procedure necessitates no explicit construction of the single-particle kinetic energy functional $T_s[n]$, although such constructions have been considered (Herring 1986).

Again, in $U[n]$ we may separate out the terms which we know, namely the potential energy due to $v(r)$ and the electrostatic energy due to the electron charge distribution

$$U[n] = \int \mathrm{d}r\, v(r)\, n(r) + \frac{1}{2}\int \mathrm{d}r \int \mathrm{d}r \int \mathrm{d}r'\, n(r) \frac{e^2}{|r-r'|} n(r') + E_{\mathrm{xc}}[n], \tag{2.6}$$

and name the remainder the exchange-correlation energy $E_{\mathrm{xc}}[n]$. The corresponding effective potential is composed of:

$$v_{\mathrm{eff}}(r) = v(r) + \int \mathrm{d}r' \frac{e^2}{|r-r'|} n(r') + v_{\mathrm{xc}}(r), \tag{2.7}$$

i.e. the external potential, the electrostatic potential due to the electronic charge distribution, and the exchange-correlation potential

$$v_{\mathrm{xc}}(r) = \delta E_{\mathrm{xc}}/\delta n(r). \tag{2.8}$$

In this way, the exchange-correlation term is unambiguously defined and at the same time the motivation of the separation, to isolate the single-particle kinetic energy, is made clear. However, without a more explicit expression for the exchange-correlation potential or energy, the term is given a certain mysterious shroud. We discuss next a formal construction.

3. A formal construction of exchange and correlation energy

It is possible to construct a perturbation series in powers of the Coulomb interaction. In the density functional theory, the unperturbed state is the non-interacting electron state which produces the same ground-state density as the exact one. The same effective potential $v_{\mathrm{eff}}(r)$ holds for the unperturbed and exact states. Then a coupling constant integral for the exchange-correlation energy functional can be derived (Harris & Jones 1974). An alternative of making the density at every intermediate coupling constant value equal to the exact density (Gunnarsson & Lundqvist 1976) simplifies the expression sufficiently to see the physical picture of an exchange-correlation hole. In actual practice of an approximate evaluation, the unknown coupling constant dependence is difficult to account for and is generally neglected without justification.

From the Harris–Jones choice of the unperturbed state, the coupling constant integral can be formally evaluated into a perturbation series (Sham 1985):

$$E_{\mathrm{xc}}[n] = \mathrm{i}\,\mathrm{Tr}\,[\ln(1-\Sigma G_0) + \Sigma G] + Y, \tag{3.1}$$

where G_0 and G are the unperturbed and exact one-electron Green's functions, Σ is the exchange-correlation part of the self-energy, trace Tr is taken over position and energy, and Y is the sum of all skeleton diagrams for the exchange-correlation energy in terms of G. Functional differentiation yields an integral equation for the exchange-correlation potential:

$$\int \mathrm{d}r'\, v_{\mathrm{xc}}(r') \left[\int \mathrm{d}\omega\, G_0(r, r'; \omega)\, G(r', r; \omega) \right] = \int \mathrm{d}r_1 \int \mathrm{d}r_2 \int \mathrm{d}\omega\, G_0(r, r_1; \omega)\, \Sigma(r_1, r_2; \omega)\, G(r_2, r; \omega). \tag{3.2}$$

Thus the exchange-correlation potential is expressed in terms of the one-particle Green's function and self-energy. To treat the magnetic systems, the density has to be extended to two spin components. The extension of the expression for v_{xc} to the spin case has been given by Ng (1989).

For a confined system, the solution of the integral equation has the correct asymptotic behaviour (Sham 1985). It is interesting that the dominant contributions for the asymptotic behaviour for an atom and for a metal surface are different. At a large distance r from an atom, the limit

$$v_{xc}(r) \sim e^2/r \tag{3.3}$$

is dominated by the exchange term (defined in the next section). At a large normal distance z outside the metal surface, the limit

$$v_{xc}(z) \sim -e^2/4z \tag{3.4}$$

comes entirely from the correlation term, specifically the surface plasmon contribution to the correlation energy.

4. Density functional exchange

To first order in the Coulomb interaction, the exchange energy is from equation (3.1)

$$E_x[n] = -\frac{1}{4}\int dr \int dr' \frac{e^2}{|r-r'|} n(r,r')\, n(r',r), \tag{4.1}$$

where the one-particle density matrix is given in terms of the density functional orbitals of equation (2.3):

$$n(r,r') = \sum_j \theta(\mu-\epsilon_j)\, \psi_j(r)\, \psi_j^*(r'), \tag{4.2}$$

summing over both spin states.

The corresponding equation of the exchange potential $v_x(r)$ is exactly the same as the local potential yielding orbitals which minimize the Hartree–Fock energy (Sharp & Horton 1953). Thus $E_x[n]$ is close to the Hartree–Fock exchange energy but is not identical to it. In the density functional theory, the exchange potential $v_x(r)$ exists on its own right rather than is a local approximation to the non-local Fock potential. Consequently, the density functional correlation energy $E_c[n]$ is not the same as the correlation energy defined as the difference between the total energy and the Hartree–Fock energy.

5. Approximations for the exchange-correlation

The approximation (Kohn & Sham 1965) which makes the density functional theory widely used is the LDA:

$$E_{xc}^{LDA} = \int dr\, \epsilon_{xc}(n(r))\, n(r), \tag{5.1}$$

i.e. in a small neighbourhood the exchange-correlation energy density is given by the corresponding term $\epsilon_{xc}(n)$ of the homogeneous electrons gas at the local density. The LDA potential is given by

$$v_{xc}^{LDA}(r) = \mu_{xc}(n(r)), \tag{5.2}$$

the exchange-correlation part of the chemical potential of the homogeneous electron gas. The guiding principle is clearly that of the Thomas–Fermi approximation. The difference lies in that the LDA does not approximate the kinetic energy term $T_s[n]$, which gives the advantage of retaining the major quantum features such as the density oscillations. In the exchange-correlation part, the LDA suffers from the same disadvantages as the Thomas–Fermi approximation, such as the lack of the gradient correction and the incorrect treatment in the decaying region where the energy is lower than the potential. The latter is largely responsible for the incorrect asymptotic behaviour of LDA in a confined system.

Because LDA is shown to account for much of the exchange-correlation effect in a wide range of inhomogeneous systems, any improvement should keep the advantage of LDA. The non-local correction to LDA proposed by Langreth & Mehl (1983) retains the LDA and adds a gradient term based on an analysis of the small wave-vector behaviour of the exchange-correlation energy. The correction appears to improve the surface energy of the jellium. In tests for small atoms where the exact results are known (Pedroza 1986), the correction improves over LDA on total energy and density. For larger atoms, Langreth & Mehl (1983) have found the density functional eigenvalues of equation (2.3) far from the exact values, but the energy differences to be in good agreement, from which they inferred that the error in the exchange-correlation potential is almost constant over the atom. For bulk silicon, the theory gives improved values for the ground-state properties, such as cohesive energy, equilibrium volume, and bulk modulus but essentially the same band gap as LDA (U. von Barth & R. Car 1987, personal communication). The band gap problem arises because the gradient correction does not yield the essential discontinuity in v_{xc}. (See the next section.)

A different approach (Gunnarsson & Jones 1980) is to construct a better exchange-correlation hole in the pair correlation function than that which reproduces the LDA for the energy functional. The pair correlation appears in the coupling constant integral for the exchange-correlation energy. The construction of the so-called weighted density approximation is made to satisfy the sum rules. In atoms, it yields better total energy than LDA but not better density nor density functional eigenvalues (Pedroza 1986). It gives much too large a surface energy for the semi-infinite jellium.

A third approach is to construct approximations based on the field theoretic expression for v_{xc}, equation (3.3). Use of the LDA for the self-energy and Green's function (Sham & Kohn 1966) recovers LDA for v_{xc}. The equation with exchange only has been solved for atoms (Aashamar *et al.* 1978). A simple form of v_{xc} which was suggested by Sharp & Horton (1953) follows from putting the electron energy ϵ_k in one of the pair of Green's functions on each side of equation (3.3) to a constant. The advantage over LDA is the correct asymptotic behaviour. The disadvantage is that it does not reduce to LDA in the slowly varying density limit which means that it does not have the well-tested aspects of LDA. A form suggested by Sham (1985) is to average the self-energy over the Fermi surface for a metal and over the highest occupied density functional orbitals in an insulator. It would reduce to LDA in the slowly varying density limit and would have the correct asymptotic limit away from the surface of a confined system. It has, however, not been numerically tested.

Solution of equation (3.3) for v_{xc} depends on the construction of a self-energy. This has been done for diamond, silicon, GaAs and AlAs (Godby *et al.* 1988). The self-energy is evaluated in the random phase approximation (RPA) in a self-consistent procedure starting with orbitals in LDA. The resultant v_{xc} is quite close to the LDA potential if the corresponding RPA is used for the homogeneous electron gas. We expect that if the self-energy were evaluated beyond RPA the resultant v_{xc} would be still close to the LDA with the accurate μ_{xc} in the electron gas given by Ceperley & Alder (1980). Why would one need the density functional theory if one has to construct the self-energy? The RPA work on v_{xc} shows that it is advantageous to start with the LDA orbitals. Hartree–Fock orbitals, for example, in general would be a poor starting approximation in solids and would in addition be much more difficult to obtain.

6. The band gap problem and discontinuity in v_{xc}

Against the wide-ranging success of LDA in ground-state properties, the uniform lack of agreement between LDA and experiment on the band gaps in semiconductors and insulators is striking. LDA values are always smaller than experiment, roughly different by a factor of two. Two possible sources of error are the LDA for v_{xc} and the use of the eigenvalues of the density functional equation (2.3) for the energies of the excited states.

In semiconductors, the v_{xc} calculated beyond LDA using RPA for the self-energy as described in the last section (Godby *et al.* 1988) yields a band gap only slightly bigger than the LDA gap. The LDA potential is sufficiently accurate that the large discrepancy in band gaps cannot come from this approximation.

In general, the eigenvalues ϵ_j from the effective one-particle equation (2.3) with the exact exchange-correlation potential v_{xc} do not represent the energies of the excited states. Although the band gap may be expressed in terms of excited state energies, it does have a special relation to the lowest energy states of $N \pm 1$ particles in the same insulator with an N-particle ground state. The highest occupied density functional eigenenergy does represent the valence band edge in the insulator or semiconductor case. The band gap of an insulator or a semiconductor can be defined precisely in terms of the ground-state energy as a function E_M of the number of particles M. If the insulating ground state has N particles, the conduction band edge is the change of the total ground-state energy when an electron is added and the valence band edge is given by the change when an electron is removed:

$$E_c = E_{N+1} - E_N, \tag{6.1}$$

$$E_v = E_N - E_{N-1}. \tag{6.2}$$

The band gap is naturally the difference:

$$E_g = E_c - E_v. \tag{6.3}$$

It is straightforward to show (Sham & Schlüter 1985) from the definition of E_v and with the help of the variational theorem that the valence band edge is given by the highest occupied density functional eigenenergy:

$$E_v = \epsilon_N. \tag{6.4}$$

Now, the Hohenberg–Kohn theorem is implicitly for a fixed number of electrons, M. Equation (2.3) and v_{xc} are implicitly defined as functions of M. Since $M = N$ gives the insulating ground state, when an electron is removed, the highest occupied state is changed by a negligible amount $[O(1/N)]$ and $v_{xc}(N-1)$ is the same as $v_{xc}(N)$ but when an electron is added across the gap, the $(N+1)$th state is very different and there is a discontinuity in v_{xc} as N is changed to $N+1$ (Sham & Schlüter 1983; Perdew & Levy 1983):

$$v_{xc}(N+1) = v_{xc}(N) + \Delta_{xc}, \tag{6.5}$$

where Δ_{xc} is independent of position. It follows that the difference between the true gap, equation (6.3) and the density functional gap given by

$$\epsilon_g = \epsilon_{N+1} - \epsilon_N, \tag{6.6}$$

is just the potential discontinuity, Δ_{xc}.

A persuasive demonstration of the importance of the discontinuity is the calculation of the discontinuity for a number of semiconductors in RPA (Godby *et al.* 1988). RPA for the Green's function using the LDA basis set (Hybertsen & Louie 1987) has been shown to give very good band structures for a number of semiconductors, including the band gaps. The density functional potential for a number of semiconductors, diamond, Si, GaAs, and AlAs, calculated (Godby *et al.* 1988) from the self-energy in RPA by iteration from LDA is then used to construct the Green's function in RPA and to determine the quasi-particle energies. The band gaps are in good agreement with experiment and the v_{xc} discontinuity accounts for a major part of the correction.

7. The Fermi surface problem

The Fermi surface problem is the counterpart of the band gap problem for conductors. The highest occupied energy in the density functional equation (2.3) can easily be shown to be the energy for adding an electron to a metal and is thus the chemical potential. The problem is whether the Fermi surface given by the eigenenergies of equations (2.3) is the same as the true Fermi surface. The true Fermi surface can be constructed from the one-particle Green's function which gives the quasi-particle energies. The LDA-like approximation for the Green's function (Sham & Kohn 1966) yields the same Fermi surface as equation (2.3). Thus, for the case of slowly varying density, the density functional Fermi surface from equation (2.3) is the same as the true Fermi surface. This includes the special case of constant density, where the isotropy of the system ensures both Fermi surfaces to be spherical and where the same total number of electrons enclosed by the Fermi surfaces guarantees the radius to be the same.

Mearns (1988) has demonstrated that the density functional Fermi surface and the true Fermi surface differ to second order in the lattice potential and first order in the Coulomb interaction. This roughly corresponds in the semiconductor case to the demonstration of the discontinuity of v_{xc} existing in the Hartree–Fock approximation (Sham & Schlüter 1985). Examples abound demonstrating that the LDA Fermi surfaces are very good approximations to the measured ones, even for hybridized f bands (Koelling 1982; Johanson *et al.* 1983). What is needed is first a demonstration of the likely event that the LDA Fermi surface is a good approximation to the density functional Fermi surface and then a theoretical understanding of the difference of the density functional Fermi surface and the true one and of the reason for the small difference.

8. Summary

For the ground-state properties, the density functional theory reduces the solution of the many-electron Schrödinger equation to that of an effective one-electron Schrödinger equation, requiring the construction of an exchange-correlation potential. The LDA offers simplicity and universality in the sense that only a knowledge of the exchange-correlation effect of the homogeneous electron gas as a function of density is needed as input. Approximations such as the inclusion of a gradient correction or a better pair correlation yield improvement in limited instances. For characteristics such as the potential discontinuity in the insulating system, construction of the self-energy for the particular system appears necessary. Simplified approximations for the self-energy (Hanke & Sham 1988) may facilitate computation of v_{xc} but their accuracy requires extensive testing. A particularly interesting direction would be to utilize the knowledge acquired from the still developing research on models of strong correlation of narrow band electrons to construct functionals for more general solids.

For quasi-particle energies, the density functional eigenvalues, especially the LDA values, form a convenient approximation though unjustified. To include the exchange-correlation effect properly, there is no avoiding the construction of the self-energy. RPA has been shown to be adequate for the band structures but not total energies in the covalent semiconductors. For metals, as in jellium, corrections to RPA are important (Hybertsen & Louie 1987). The LDA orbitals form a good starting basis for the construction of the self-energy but it appears that to improve on LDA a simple universal approximation may no longer be adequate.

I acknowledge the benefit of helpful discussions from many persons, especially my collaborators, R. Godby, W. Kohn and M. Schlüter, over the years. This work is supported in part by NSF Grant no. DMR 88-15068.

References

Aashamar, K., Luke, T. M. & Talman, J. D. 1978 *At. Data Nucl. Data Tables* **22**, 443–472.

Ceperley, D. M. & Alder, B. J. 1980 Ground state of the electron gas by a stochastic method. *Phys. Rev. Lett.* **45**, 566–569.

Godby, R. W., Schlüter, M. & Sham, L. J. 1988 Self-energy operators and exchange-correlation potentials in semiconductors. *Phys. Rev.* **37**, 10159–10175.

Gunnarsson, O. & Jones, R. O. 1980 *Physica Scr.* **21**, 394–401.

Gunnarsson, O. & Lundqvist, B. I. 1976 Exchange and correlation in atoms, molecules, and solids by the spin-density-functional formalism. *Phys. Rev.* B **13**, 4274–4298.

Hanke, W. & Sham, L. J. 1988 Density-functional theory in insulators: Analytical model for Σ_{xc}, v_{xc}, and the gap correction. *Phys. Rev.* B **38**, 13361–13370.

Harris, J. & Jones, R. O. 1974 *J. Phys.* F **4**, 1170–1186.

Herring, C. 1986 Explicit estimation of ground-state kinetic energies from electron densities. *Phys. Rev.* A **34**, 2614–2631.

Hybertsen, M. S. & Louie, S. G. 1987 Theory and calculation of quasi-particle energies and band gaps. *Comments Cond. Matt. Phys.* **13**, 223–248.

Hohenberg, P. & Kohn, W. 1964 Inhomogeneous electron gas. *Phys. Rev.* **136**, B864–871.

Johanson, W. R., Crabtree, G. W., Edelstein, A. S & McMaster, O. D. 1983 *J. Magn. Magn. Mat.* **31–34**, 377–381.

Koelling, D. D. 1982 *Solid St. Commun.* **43**, 247–251.

Kohn, W. & Sham, L. J. 1965 Self-consistent equations including exchange and correlation effects. *Phys. Rev.* **140**, A1133–1138.

Langreth, D. C. & Mehl, M. J. 1983 *Phys. Rev.* B **28**, 1809–1834.

Lundqvist, S. & March, N. H. (eds) 1983 *Theory of the inhomogeneous electron gas.* (395 pages.) New York: Plenum.

Mearns, D. 1988 Inequivalence of the physical and Kohn–Sham Fermi surfaces. *Phys. Rev.* B **38**, 5906–5912.

Ng, T. K. 1989 Exchange-correlation potentials in density- and spin-density-functional theory. *Phys. Rev.* B **39**, 9947–9958.

Pedroza, A. C. 1986 Nonlocal density functionals: comparisons with exact results for finite systems. *Phys. Rev.* A **33**, 804–813.

Parr, R. G. & Yang, W.-T. 1989 *Density-functional theory of atoms and molecules.* (333 pages.) New York: Oxford University Press.

Perdew, J. P. & Levy, M. 1983 Physical content of the exact Kohn–Sham orbital energies. *Phys. Rev. Lett.* **51**, 1884–1887.

Schlüter, M. & Sham, L. J. 1982 Density functional theory. *Physics Today* **35**, 36–43.

Sham, L. J. 1985 Exchange and correlation in density-functional theory. *Phys. Rev.* **32**, 3876–3882.

Sham, L. J. & Kohn, W. 1966 One-particle properties of an inhomogeneous interacting electron gas. *Phys. Rev.* **145**, 561–567.

Sham, L. J. & Schlüter, M. 1983 Density-functional theory of the energy gap. *Phys. Rev. Lett.* **51**, 1888–1891.

Sham, L. J. & Schlüter, M. 1985 Density-functional theory of the band gap. *Phys. Rev.* **32**, 3883–3889.

Sharp, R. T. & Horton, G. K. 1953 *Phys. Rev.* **90**, 317.

Discussion

Z. Reut (*City University, London, U.K.*). It may be more appropriate to define the psuedo- or effective potential by an integral, rather than by the usual gradient that might not exist, or be of a dubious meaning in the case of many-electron systems. The effective potential in such cases can be velocity dependent and subject to relativistic effects.

L. J. Sham. For a physical system, small changes of the electron density generally produce small changes in $E_{xc}[n]$ (except across the band gap of a semiconductor). Therefore, $\delta E_{xc}[n]/\delta n$ is well defined. For a semiconductor, it is well defined for either side of the hyperface of the constant total electron number.

J. W. Wilkins (*Ohio State University, U.S.A.*). In strained superlattices – such as GaAs/GaInAs grown in the [III] direction – there are large, strain-induced, electric fields of order 10^5 V cm^{-1} or 1 meV Å^{-1}. Can density functional theory adequately explain excitations between the valence and conduction bands in such a situation?

L. J. Sham. The band structure computation itself at the present stage is unable to reproduce accurately the band edge features in the meV range. It is, therefore, necessary to use the effective mass theory with the relevant band edge position and curvature taken from experiment.

V. Heine (*Cambridge, U.K.*). Professor Sham emphasized the corrections beyond the LDA to band gaps, where there are experimental data to compare with. That is natural, but I wonder whether in the long run some chemical effects may not be more important. For example in a transition metal, the exchange and correlation hole seen by a d electron will be quite different from the hole seen by an sp electron, unlike the situation in LDA. Thus the relative position of the sp and d band will be incorrect, which in turn affects the amount of hybridization. Similarly in high T_c superconductors, much hangs on the balance between the Cu 3d level and the oxygen 2p level, which one cannot expect to give well by the LDA.

L. J. Sham. For the states close to the Fermi level, the density functional theory (DFT) gives quite good Fermi surface even though not exact. So improvement from the construction of the DFT self-energy should be small. But for deeper levels, such as the sp and d band separation, the construction of the one-particle Green's function is necessary.

L. M. FALICOV (*University of California, Berkeley, U.S.A.*). For moderately correlated systems (ferromagnets, antiferromagnets) the DFT gives qualitatively the correct Fermi surface only if the correct symmetry breaking is considered from the start. For highly correlated heavy fermions, where symmetry breaking normally takes place and is much more subtle, why does Professor Sham 'hope' to obtain the correct Fermi surface?

L. J. SHAM. LDA has given Fermi surfaces even in heavy Fermion metals quite close to the measured ones. One needs to understand the reason.

D. M. EDWARDS (*Imperial College, London, U.K.*). Professor Sham discussed the formulations of the exact exchange correlation potential using perturbation theory. Since even the exact functional does not give the exact Fermi surface isn't there a problem with the anomalous diagrams of Kohn and Luttinger?

L. J. SHAM. No, in the perturbation expansion starting from the density functional Green's function G_0, one can avoid the anomalous diagrams by keeping the unperturbed and perturbed Fermi levels the same, which they naturally are in DFT.

Many-body effects in layered systems

By J. A. White† and J. C. Inkson
University of Exeter, Department of Physics, Stocker Road, Exeter EX4 4QL, U.K.

An investigation of quasi-particle properties is presented in single- and double-quantum-well systems as a function of electron density. Significant changes from bulk parameters are found with strong dependencies on the electron density and interwell separation. Low-energy plasmon modes are found to dominate the quasi-particle lifetime giving scattering rates up to an order of magnitude higher than optic phonons in these systems.

1. Introduction

The electronic properties of single and multiple layer quasi-two-dimensional systems have been studied extensively both experimentally and theoretically in recent years driven by both the technological importance and applications of the new semiconductor growth systems and by the appearance of new physics (Ando *et al.* 1982). Reduced dimensionality, changes in coupling to phonon modes, the effects of interface scattering, strain have all been considered. The study of many-body effects in two-dimensional systems is particularly interesting because, apart from the usual structural parameter variations available, the two-dimensional electron density can be varied continuously over a wide range in a single device and so provides the opportunity to test the theory in much more detail than is possible in bulk materials.

Many-body effects have been studied in silicon inversion layers (Ando *et al.* 1982; Smith & Stiles 1972; Ting *et al.* 1975; Ohkawa 1976; Vinter 1976) but recently, GaAs or AlAs quantum well systems have become the centre of attention. In such systems, because of the low GaAs band mass, many-body effects are not expected to be as important as in silicon. There are still significant exchange-correlation contributions to the subband structure (Kawamoto *et al.* 1980; Ando 1987), however, while the absence of extraneous intervalley and impurity scattering and the direct nature of the band gap have important theoretical and experimental advantages.

The central quantity in the calculation of quasi-particle properties is the electron self-energy. This requires the calculation of the screened electron–electron interaction which in turn involves the calculation and inversion of the dielectric response function the poles of which are the plasmons. Physically the self-energy can be thought of as arising largely from the virtual emission and absorption of plasmons with the other (single-particle excitations) playing a minor role and so the study of interaction effects is linked closely to that of plasmon properties.

There have been many calculations of the plasmon spectrum in single and multiple layers (Stern 1967; Dahl & Sham 1977; Eguiluz & Maradudin 1978; Tselis & Quinn 1982) up to superlattice systems (Grecu 1973; Hawrylak *et al.* 1985; King-Smith & Inkson 1986), i.e. an infinite array of parallel two-dimensional electron gases of equal

† Present address: Department of Theoretical Physics, University of Lund, Sölvegatan 14A, S-22362 Lund, Sweden.

electron density. The conclusion for the single-quantum-well system with one occupied band is that there are as many plasmon bands as there are electronic subbands. The lowest plasmon band being due to intraband and the higher bands interband polarization. Multiple-well systems support acoustic plasmon modes through the interlayer interaction, the number of such modes increasing with the number of layers. These plasmon modes have been measured in light scattering experiments (Olego *et al.* 1982; Fasol *et al.* 1986) and the agreement between theory and experiment has been good.

The energies of the plasmons are typically of the order of both the intersubband energies and, for typical experimental parameters, optical phonon energies. This together with the reduced dimensionality suggests that the self-energy and hence the quasi-particle parameters can be expected to show much more structure than in the case of bulk semiconductors. Few such calculations have been carried out, however, Hawrylak and co-workers (Hawrylak 1987; Hawrylak *et al.* 1988) have calculated the correlation energy, effective mass and quasi-particle lifetime within the RPA for quasi-particles in a superlattice. Here we shall concentrate on the properties of a single quantum well and two parallel quantum wells so that low-dimensional effects can be emphasized. These systems also offer the easiest possibility of experimentally varying the electron density within the same device.

2. Model

We first take the case of a single well; we consider the lowest two subbands, $n = 0, 1$, of a single modulation doped semiconductor quantum well of width a within the effective mass approximation. For simplicity we assume that only the lowest subband is occupied with an areal electron density N. We also assume that the subband wavefunctions are those of an infinite square well. This is expected to be a good enough approximation for symmetric quantum wells in the GaAs or GaAlAs system when the barrier height is high compared with the energy gap between the subbands. Electrons exhibit free particle-like behaviour in the plane of the well, with 'bare' mass m for each subband. The screening of the Coulomb interaction by the intrinsic semiconductor is described by a dielectric constant κ.

The hamiltonian for the system then is that of particles in a simple two-dimensional square well potential interacting via a Coulomb interaction $e^2/\kappa r$. The spatial part of the wavefunction for the single-particle state with parallel momentum $\boldsymbol{k}$, in subband n is thus

$$\psi_{\boldsymbol{k}}^{n}(\boldsymbol{r}) = \phi_n(z)\,\mathrm{e}^{\mathrm{i}\boldsymbol{k}\cdot\boldsymbol{\rho}},$$

where $\boldsymbol{r} = (\boldsymbol{\rho}, z)$ and any $\boldsymbol{k}$ dependence of the $\phi_n(z)$ is neglected. Denoting the corresponding annihilation operator for spin σ as $c_{kn\sigma}$, the hamiltonian may be written

$$H = \sum_{qn\sigma} t_{qn}\, c^{+}_{qn\sigma} c_{qn\sigma} + \frac{1}{2} \sum_{\substack{q\neq 0\\ kk'\sigma\sigma'\\ ll'nn'}} V_q^{ll'nn'}\, c^{+}_{k+ql\sigma}\, c^{+}_{k'-qn\sigma'} c_{k'n'\sigma'}\, c_{kl'\sigma},$$

where $t_{qn} = E_n + q^2/2m$ are the 'single-particle' nth subband energies, with minima at $E_0 = 0$ and $E_1 = E_{\mathrm{g}}$.

$$V_q^{ll'nn'} = \frac{2\pi}{\kappa q} \int\!\!\int \mathrm{d}x\,\mathrm{d}x'\, \phi_l(x)\,\phi_{l'}(x)\,\mathrm{e}^{-q|x-x'|}\phi_n(x')\,\phi_{n'}(x') \equiv v_q\, D_q^{ll'nn'}$$

are the Coulomb matrix elements between subband wavefunctions $ll'nn'$ for a momentum exchange of $\hbar q$, $v_q = 2\pi/\kappa q$ the two-dimensional Fourier transform of the Coulomb interaction and D is the weighting factor. All other symbols have their usual meaning. In the hamiltonian (2) we have included all electron–electron interactions beyond the Hartree approximation in the second term.

From the orthonormality of the subband wavefunctions it is easy to show that, D^{0000}, D^{1111}, $D^{0011} \rightarrow 1$ and $D^{0101} \propto qa$ as $qa \rightarrow 0$, while all of these four terms are proportional to $1/qa$ for $qa \rightarrow \infty$. By symmetry we also have $D^{0001} \equiv D^{1110} \equiv 0$.

The RPA dielectric function for the system is most easily dealt with by considering the Fourier transform in the plane and then expressing it as a matrix in the subband pairs appropriate to the Coulomb matrix elements in equation (3), thus:

$$\epsilon_{\alpha\beta}(q, \omega) = \delta_{\alpha\beta} - V_q^{\alpha\beta} P_\beta(q, \omega),$$

where the index α, β are the subband pairs (00), (01), (11). We shall denote such subband pairs with Greek letters throughout this paper.

The calculation for the intrasubband polarization P_{00} has been performed by Stern (1967) and for the intersubband term P_{01} we use the formulation of King-Smith and Inkson (1986). Since only the lowest subband is occupied, $P_{11} \equiv 0$. The spectrum of excitations is given by the inverse dielectric matrix

$$\mathrm{Im}\, \epsilon_{\alpha\beta}^{-1}(q, \omega) \neq 0,$$

i.e. intraband electron–hole pairs when $\mathrm{Im}\,(P_{00}(q, \omega)) \neq 0$, interband electron–hole pairs when $\mathrm{Im}\,(P_{01}(q, \omega)) \neq 0$ and plasmons when $\det \epsilon_{\alpha\beta}(q, \omega) = 0$.

The intraband plasmon corresponds to charge density oscillations in the plane of the quantum well and has the well-known $\omega \propto q^{\frac{1}{2}}$ behaviour at long wavelengths. The interband plasmon corresponds to charge density oscillations across the well and its energy tends to a constant value in the $q \rightarrow 0$ limit. The excitation spectrum is shown in figure 1 for GaAs parameters ($m = 0.07$, $\kappa = 13$) when well width $a = 100$ Å†, $N = 3.3 \times 10^{11}\ \mathrm{cm}^{-2}$, and the subband gap has been chosen as 30 meV. Also note that, by symmetry, there is no Landau damping of the intraband plasmons lying within the interband electron–hole continuum.

In the case of a double well, the effects we are concerned with arise from the coupling of the low-energy intraband plasmons. The model is, therefore, simpler and includes only a single subband in each well. We consider a system of two parallel two-dimensional electron gases, labelled $i = 1, 2$, with areal electron densities n_1, n_2 and with interlayer separation d, i.e. electrons are confined to move in the planes $z_1 = 0$, $z_2 = d$ with no tunnelling between the layers. The in-plane effective 'bare' mass is m and the background dielectric constant is κ. This model is sufficient to describe fairly accurately the electronic response properties of a doped double-quantum-well structure when only the lowest subband in each well is occupied. It neglects the finite width of the wells and also neglects the effect of intersubband transitions. The effect of finite well width could be included through the usual form factors for the Coulomb interaction. This reduces the short-range part of the interaction, but should produce, qualitatively, the same results.

The dielectric function becomes a 2×2 matrix in real space

$$\epsilon_{ij}(q, \omega) = \delta_{ij} - V_{ij}(q) P_j(q, \omega), \quad i, j = \text{layer } 1, 2,$$

† 1 Å $= 10^{-10}$ m.

Figure 1

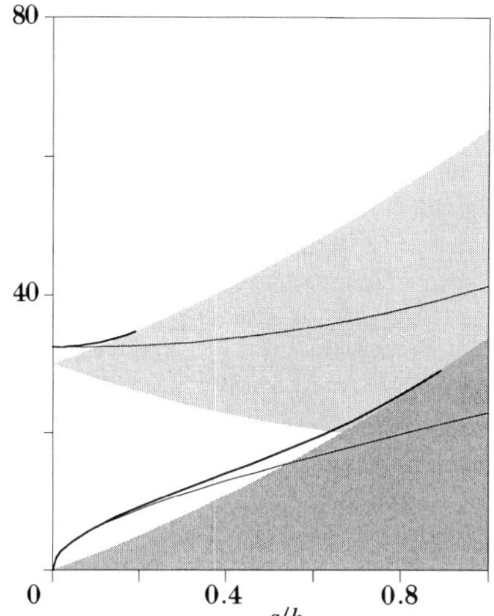

Figure 2

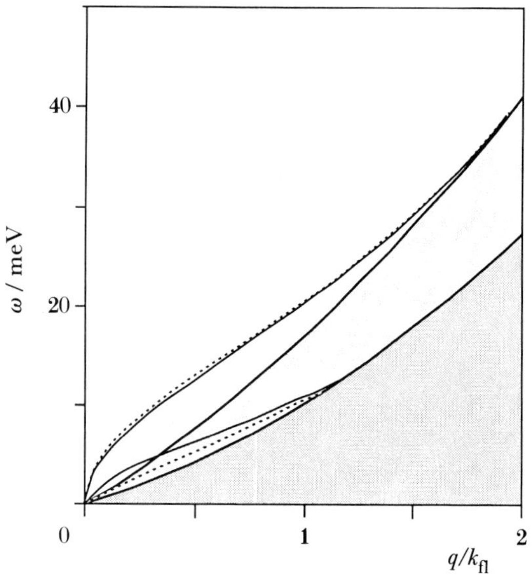

Figure 1. The plasmon modes (thick lines) of a doped quantum well with $n = 3.3 \times 10^{11}$ cm^{-2}, well width $a = 100$ Å, $E_g = 300$ meV, $\kappa = 13$, and $m = 0.07$. The shaded areas show the intraband and interband electron–hole continua. Also shown (thin lines) are the effective intraband and interband plasmon poles of equation (19), which approximate the plasmons for small q and lie inside the electron–hole pair bands for large q.

Figure 2. Plasmon spectrum for a two layer system with $n_1 = 10^{11}$ cm^{-2} ($k_{\mathrm{f1}} = 8 \times 10^5$ cm^{-1}) and $n_2 = 4 \times 10^{11}$ cm^{-2} ($k_{\mathrm{f2}} = 1.6 \times 10^6$ cm^{-1}) for $d = 100$ Å (solid lines) and for $d = 1000$ Å (broken lines). The shaded areas indicate the electron-hole continua of each layer. Plasmons in the shaded area are Landau damped as they may excite electron–hole pairs in the low density layer.

where q is the magnitude of the in-plane momentum, $P_i(q, \omega)$ is the polarization for layer i, and $V_{ij}(q)$ is the Coulomb interaction between layers i and j.

The screened interaction between layers i and j is then

$$W_{ij}(q, \omega) = \sum_l \epsilon_{il}^{-1}(q, \omega)\, V_{lj}(q)$$

and the poles of the screened interaction occur when

$$\det \epsilon_{ij}(q, \omega) = 0.$$

Unless the densities in the two layers are identical, no analytic solutions for the plasmon dispersion can, in general, be found. However, in the long wavelength limit, the high-frequency mode is obtained as just the $\omega_+ \propto q^{\frac{1}{2}}$ dependence of a plasmon in a single two-dimensional EG of density $(n_1 + n_2)$ and our numerical studies also show that the low-energy mode has the acoustic form $\omega_- \propto q$ at long wavelengths with the plasmon velocity decreasing as d decreases. Figure 2 shows the typical plasmon dispersion relations when $n_1 = 10^{11}$ cm^{-2}, $n_2 = 4 \times 10^{11}$ cm^{-2} for two separations, $d = 100$ Å and $d = 1000$ Å.

For the larger separation ω_- lies outside the electron–hole continua of both layers for small q and is therefore, like the ω_+ mode, undamped. For larger q values this mode lies inside the electron–hole pair band of layer 2 and is Landau damped, i.e. the charge density oscillation may loose energy by exciting electron–hole pairs in layer 2. This damping is very weak, however.

3. Quasi-particle properties

For the single layer, by symmetry, the self-energy is diagonal in the subband index and each of the diagonal elements has two contributions which we label by the band index of the internal Green function

$$\Sigma_{nm}(k,E) = \frac{\mathrm{i}}{8\pi^3}\sum_l \int G^0_{ll}(\boldsymbol{k}+\boldsymbol{q},E+\omega)\, W_{nllm}(q,\omega)\,\mathrm{e}^{\mathrm{i}\omega\delta}\,\mathrm{d}\boldsymbol{q}\,\mathrm{d}\omega$$

$$\equiv \sum_l \Sigma^l_{nn}(k,E)\,\delta_{nm},$$

where G^0_{nm} is the non-interacting Green function for electrons in subband n and the $W_{ll'nn'}$ are matrix elements of the screened interaction

$$W_{\alpha\beta}(q,\omega) = \sum_\gamma \epsilon^{-1}_{\alpha\gamma}(q,\omega)\, V^{\gamma\beta}_q.$$

Physically, Σ^n_{nn} corresponds to the virtual exchange of an intraband plasmon, the electron remaining within the subband whilst $\Sigma^{n'}_{nn}$ involves an interband plasmon and an intermediate state in the alternate subband.

The quasi-particle energies are then given approximately by

$$E_n(k) = t_{kn} + \Sigma_{nn}(k, t_{kn}).$$

For the double layer, the self-energy is calculated within the GW approximation on the mass shell. As there is no electron tunnelling between the layers, the self-energy is now diagonal in the layer index, i.e. it depends only on the effective intralayer interaction.

$$\Sigma_{mn}(k, E_0(k)) = \frac{\mathrm{i}\delta_{mn}}{8\pi^3}\int G_{mm}(\boldsymbol{k}+\boldsymbol{q}, E_0(k)+\omega)\, W_{mm}(q,\omega)\,\mathrm{d}\boldsymbol{q}\,\mathrm{d}\omega.$$

These relations are evaluated numerically to give the self-energy and hence the quasi-particle properties.

Figure 3 shows, for a single well, the real part of the self-energy at the quasi-particle peak, i.e. the contributions to the effective exchange correlation potentials as a function of parallel momentum.

First consider the two largest terms, Σ^0_{00} and Σ^1_{11}, which involve intraband (virtual) scattering processes. The self-energy in the lower subband has, generally, the larger magnitude. This means that, the effect of exchange and correlation is to increase the subband gap. The other significant feature of these two terms is the sharp downward 'spike' at around $k = 1.4k_{\mathrm{f}}$ which occurs at the threshold for an electron to emit plasmons. This is seen in three-dimensional systems as well (Lundqvist 1967; Overhauser 1971), but is more pronounced here because of the reduced dimensionality and incomplete screening. This 'spike' is largest in the higher subband because the electron–plasmon scattering is restricted in the lower subband by the Pauli principle.

The terms Σ^1_{00} and Σ^0_{11} involve intersubband scattering and are much smaller than the intraband scattering contributions because of the smaller Coulomb matrix

Figure 3

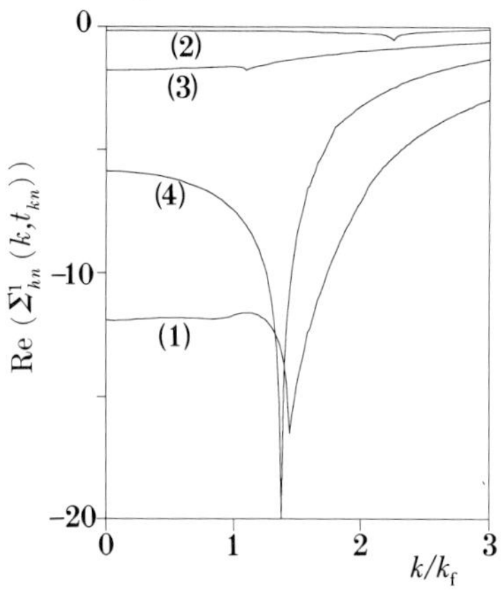

Figure 4

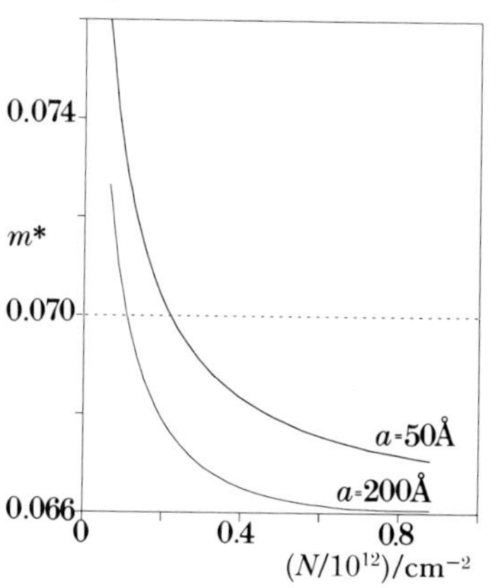

Figure 3. The real part of the contributions to the self-energy, evaluated at the quasi-particle peak, as a function of k. Σ^0_{00} (line (1)), Σ^1_{00} (line (2)), Σ^0_{11} (line (3)), and Σ^1_{11} (line (4)), as defined by equation (22), are shown. Here $N = 6.4 \times 10^{11}$ cm^{-2}, $E_g = 30$ meV and $a = 100$ Å.

Figure 4. Quasi-particle effective mass m^* as a function of electron density for well widths $a = 50$ Å and $a = 200$ Å.

elements. Again, each term has a spike corresponding to the threshold for emission of the effective intersubband plasmons. These are just visible at $k \approx 1.1k_f$ for Σ^0_{11} and $k \approx 2.3k_f$ for Σ^1_{00}.

3.1. *Effective masses*

The quasi-particle effective mass m^*, defined in terms of the quasi-particle group velocity v_g at the Fermi surface, is given by

$$\frac{1}{m^*} = \frac{1}{m} + \frac{1}{k_f}\frac{d}{dk} \operatorname{Re} \Sigma_{00}(k, t_{k0})\,|_{k=k_f}.$$

This can easily be calculated to the desired accuracy by numerical differentiation of the self-energy. We find that the behaviour of m^* is primarily determined by the intraband term in the self-energy Σ^0_{00}. Figure 4 shows m^* as a function of electron density N for two values of well width, $a = 50$ Å and $a = 200$ Å. Notice that m^* is significantly lower for the wider well. This is because the short-range part of the interaction is reduced in wide wells, which increases the importance of exchange effects (which tend to reduce m^*) while reducing the electron–plasmon coupling strength.

From figure 3 it is clear that there could be much more significant changes in the effective mass due to the very strong structure in the self-energy. However, for a single well, like a bulk material, the Fermi level lies well away from the region, corresponding to the plasmon emission threshold, in which the self-energy is varying rapidly. Lowering that threshold towards the Fermi energy could be expected to produce large changes in the effective mass. The most effective way of doing this is to introduce a second well. Coupling between the plasmons in the wells will introduce a lower energy 'acoustic' plasmon without changing the Fermi level. A suitable choice of densities and structural parameters also give the possibility of matching the plasmon velocity to the Fermi velocity so that resonant effects might occur.

Figure 5 shows the resulting variation in the Fermi level effective mass for a range of electron densities in the second well for a fixed separation. There is a wide range

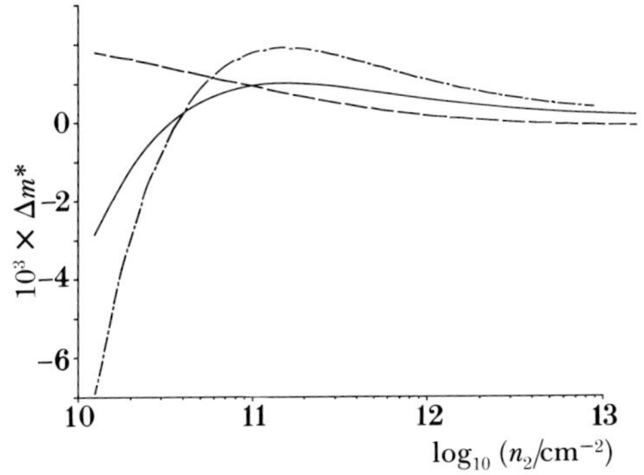

Figure 5. Difference Δm^* between effective mass at $d = 62.5$ Å and its value in a single two-dimensional EG ($d \to \infty$) as a function of density n_2 with $n_1 = n_2$ (solid curve) and $n_1 = 10^{11}$ cm^{-2} (broken curve). Also for the superlattice with $n = n_2$ (chain curve).

of behaviour with mass increases and decreases. This is to be expected since there are a number of effects in operation; the coupling to the higher plasmon mode, the acoustic plasmon velocity and the conservation of total oscillator strength. In particular the lower the energy of the acoustic plasmon with respect to the single-well energy, the lower its oscillator strength which reduces the overall effect. However, it is clear that this system would be a rich one for experimental investigation.

3.2. *Quasi-particle lifetime*

In a quantum well system the quasi-particle lifetime is governed by the possibility of the excitation of plasmons, phonons or electron–hole pairs. In our present calculations phonons are neglected and in the plasmon pole approximation used for most of these calculations, the electron–hole excitations have been subsumed within the plasmon oscillator strength. Calculations using the full RPA response function show the effect of the latter to be small anyway and what we are concerned with here is the magnitude of the lifetime, which is dominated by plasmon emission.

The inverse lifetime for quasi-particles of momentum k in subband n due to scattering into subband n' is

$$\tau^{-1}_{knn'} = -2\,\mathrm{Im}\,\Sigma^{n'}_{nn}(k, t_{kn}). \tag{26}$$

Results for the inverse lifetimes due to intrasubband scattering as a function of quasi-particle momentum k, τ^{-1}_{k00} and τ^{-1}_{k11}, are shown in figure 6 for an electron density $N = 6.4 \times 10^{11}$ cm^{-2} and for well widths $a = 50$ Å and $a = 200$ Å. The inverse lifetime has been plotted in units of energy, with 6.56 meV corresponding to 10^{13} s^{-1}. Since the Coulomb matrix elements for intraband scattering are reduced as the well width increases, the larger well width corresponds to a lower emission threshold and also to a reduced scattering rate. The discontinuous change in the intraband scattering rate in the higher subband is typical of plasmons emission in a two-dimensional electron gas (Hawrylak *et al.* 1988). The discontinuity is not seen for intraband scattering within the lowest subband because of the restrictions of the exclusion principle on scattering.

Figure 7 shows the lifetime for intraband scattering for a double well for a range of separations. Both coupled plasmon modes contribute significantly for all separations. As the separation decreases, however, the range over which an acoustic dispersion holds increases. This is seen in the lifetime as both a large shift and a

Figure 6

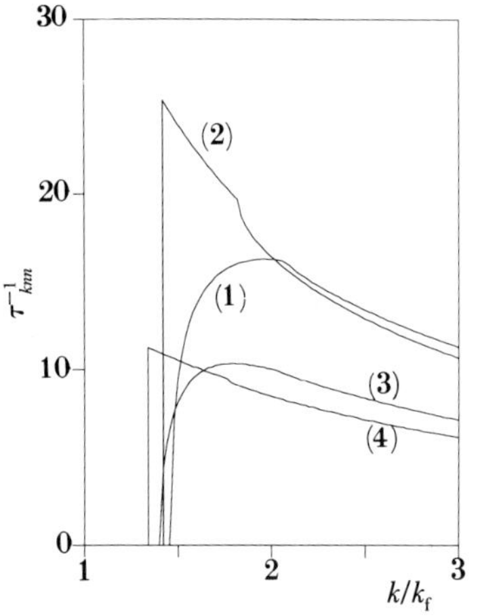

Figure 7

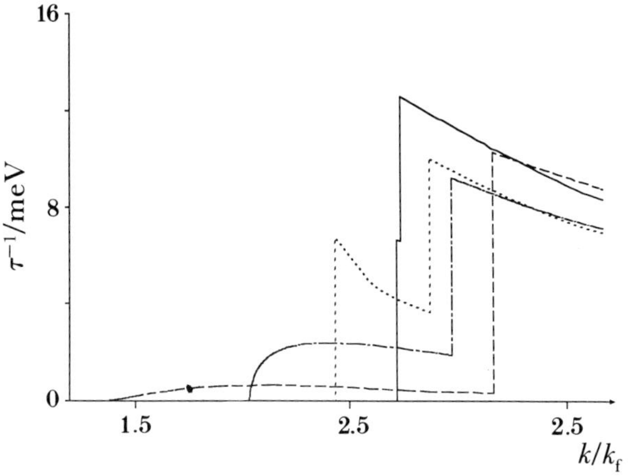

Figure 6. Inverse lifetime τ^{-1}_{knn} to intrasubband scattering within subband n as a function of quasiparticle parallel momentum k. Plotted for both subbands and for two well widths: $n = 0$, $a = 50$ Å (line (1)); $n = 1$, $a = 50$ Å (line (2)); $n = 0$, $a = 200$ Å (line (3)); $n = 1$, $a = 200$ Å (line (4)). Here $N = 6.4 \times 10^{11}$ cm^{-2} and $E_g = 30$ meV.

Figure 7. Inverse lifetime to plasmon emission in a two layer system, $n_1 = n_2 = 10^{11}$ cm^{-2}, for interlayer separations: 800 Å (solid), 300 Å (dotted), 200 Å (chain), 100 Å (broken).

functional change (to $(k-k_t)^{1.5}$) in the threshold, corresponding physically to the increasing preponderance of small-angle scattering allowed. (The equivalent superlattice threshold by contrast has an exponent of 2.)

For the electron densities considered here, the rate for intrasubband scattering via the Coulomb interaction is large compared with other scattering processes, such as scattering from LO phonons (Mason & Das Sarma, 1987). Furthermore intraband plasmon emission generally corresponds to large momentum transfers (*ca.* k_f) so that it is a very effective channel for momentum relaxation. It is therefore expected to be the dominant scattering process in hot electron transport for either single or coupled wells. The intersubband scattering rate, although much lower than the intrasubband rate (*ca* 10%), is still significant. For $a = 200$ Å, the intersubband lifetime of about 10^{-12} s is still similar to that for interband relaxation by LO phonon emission (Ridley 1989; Tatham *et al.* 1989), however. It might therefore be expected that electron–electron interactions will become the dominant intrasubband relaxation process in wide wells for which the subband gap is small enough to restrict LO phonon emission.

These Coulomb scattering rates, both intraband and interband, increase as electron density increases. We find that the intraband scattered rate scales approximately as between $N^{\frac{1}{2}}$ and N, while the interband scattering rate scales approximately as N.

4. Conclusions

We have shown that the modification of the plasmon modes in layered systems can alter basic quasi-particle properties. These properties are sensitive to both the structure and electron densities involved and are therefore a prime area for the study of electron–electron interactions, plasmon modes and their influence upon the quasi-particle properties.

References

Ando, T. 1987 Theory of semiconductor heterostructures. In *Third Brazilian school of semiconductor physics* (ed. C. E. T. Gonçalves da Silva, L. E. Oliveira & J. R. Leite), pp. 23–45. World Scientific.

Ando, T., Fowler, A. B. & Stern, F. 1982 Electron properties of two dimensional systems. *Rev. mod. Phys.* **54**, 437–677.

Dahl, D. A. & Sham, L. J. 1977 Electrodynamics of quasi two dimensional systems. *Phys. Rev.* B**16**, 651–659.

Eguiluz, A. & Maradudin, A. A. 1978 Electromagnetic modes of an inversion layer on a semiconductor surface. *Ann. Phys.* **113**, 29–78.

Fasol, G., Hughes, H. P. & Ploog, K. 1986 Raman scattering by coupled layer plasmons and in plane two-dimensional single particle excitations in multi quantum well structures. *Surf. Sci.* **170**, 497–513.

Grecu, D. 1973 Plasmons in layered systems. *Phys. Rev.* B**8**, 1958–1967.

Hawrylak, P. 1987 Effective mass and lifetime in a layered electron gas. *Phys. Rev. Lett.* **59**, 485–488.

Hawrylak, P., Eliasson, G. & Quinn, J. J. 1985 Electron self energy, effective mass and lifetime in a layered electron gas. *Phys. Rev.* B**37**, 10187–10203.

Hawrylak, P., Ji-Wei Wu & Quinn, J. J. 1985 Intersubband collective excitations at the surface of a semiconductor superlattice. *Phys Rev.* B**32**, 5169–5182.

Kawamoto, G., Kalia, R. & Quinn, J. J. 1980 Exchange and correlation insemiconductor surface inversion layers. *Surf. Sci.* **98**, 589–598.

King-Smith, R. D. & Inkson, J. C. 1986 Real space inversion of the dielectric response function of a superlattice. *Phys. Rev.* B**33**, 5489–5493.

Lundqvist, B. I. 1967 Single particle spectrum of an electron gas. *Phys. Kondens. Mater.* **6**, 193–218.

Mason, B. A. & Das Sarma, S. 1987 Theory of polar scattering in semiconductor quantum structures. *Phys. Rev.* B**35**, 3890–3898.

Ohkawa, F. J. 1976 Quasiparticle in surface quantised states in silicon. *Surf. Sci.* **58**, 326–332.

Olego, D., Pinczuk, A., Gossard, A. C. & Wiegmann, W. 1982 Plasmon dispersion in a layered electron gas. *Phys. Rev.* B**26**, 7867–7879.

Overhauser, A. W. 1971 Simplified theory of electron correlations in metals. *Phys. Rev.* B**3**, 1888–1898.

Ridley, B. K. 1989 Electron scattering by confined LO polar phonons in a quantum well. *Phys. Rev.* B**39**, 5282–5291.

Smith, J. L. & Stiles, P. J. 1972 Electron–electron interaction continuously variable in the range $2.1 > r_s > 0.9$. *Phys. Rev. Lett.* **29**, 102–104.

Stern, F. 1967 Polarisability of a two dimensional electron gas. *Phys. Rev. Lett.* **18**, 546–548.

Tatham, M. C., Ryan, J. F. & Foxon, C. T. 1989 Time resolved Raman measurements of intersubband relaxation in GaAs quantum wells. *Phys. Rev. Lett.* **63**, 1637–1641.

Ting, C. S., Lee, T. K. & Quinn, J. J. 1975 Effective mass and g factors of interacting electrons in the surface inversion layer of silicon. *Phys. Rev. Lett.* **34**, 870–874.

Tselis, A. & Quinn, J. J. 1982 Collective modes of semiconducting space charge layers. *Surf. Sci.* **113**, 362–381.

Vinter, B. 1976 Correlation energy and effective mass of electrons in an inversion layer. *Phys. Rev.* B**13**, 4447–4458.

Discussion

J. W. Wilkins (*Ohio State University, U.S.A.*) How would the large internal strains and resulting electric fields effect screening in superlattices?

J. Inkson. The electric fields resulting from strain or other sources (applied fields, space charge, etc.) redistribute the electron density within the ground state. This can change the quantum well potential profiles and the single-particle wavefunctions dramatically even in extreme cases splitting the quantum well electron density into two separate two-dimensional electron gases. The results of our treatment are, however, dependent primarily upon gross features such as the electron density in the layers and the orthogonality of states in differing subbands.

M. L. COHEN (*University of California, Berkeley, U.S.A.*). What about phonon effects? Shouldn't there be phonon–plasmon coupling and other effects since the energy ranges are similar?

J. INKSON. Yes of course; one of the interesting aspects of this system is precisely the similarity of the energy ranges. Farol *et al.* (1986) may already have seen phonon effect in their data. Our results show the plasma coupling to be much stronger than phonon but we are already working on incorporating the phonon–plasmon coupling into the problem.

M. J. KELLY (*GEC and Cavendish Laboratory, U.K.*). What is the probability of phonon emission by a 300 meV hot electron passing at right angles through a two-dimensional electron gas?

Additional reference

Farol, G., King-Smith, R. D., Richards, D., Ekenberg, U., Mestres, N. & Ploog, K. 1986 *Phys. Rev.* B**39**, 12695–12703.

Predicting new solids and their properties

By Marvin L. Cohen
Department of Physics, University of California, Materials Sciences Division, Lawrence Berkeley Laboratory, Berkeley, California 94720, U.S.A.

It is now possible to predict properties of materials using quantum theory and minimal information about the constituent atoms of a solid. Accurate calculations explaining and, in some cases, predicting electronic, structural, vibrational, and even superconducting properties of solids have been performed. Here we provide an overview of this area with emphasis on the predictive capacity of the approach. A few of the applications which will be highlighted include the analysis of high-pressure structures of solid, compounds having very low compressibilities, and metallic hydrogen.

1. Introduction

The theory of bonding and structure of solids has evolved considerably in the past 30 years. The quantum theory developed in the 1920s provided the fundamental tool for exploring solids, but the invention of appropriate models and the mechanics for solving the resulting differential equations were required before applications to real systems could be made. At first, considerable physical insight was achieved with simple models. Later, the advent of high-speed computers allowed the use of more appropriate complex models. Today computer modelling of materials is a successful industry and interesting advances are being made at a rapid rate (Cohen 1986). An early worry that computer modelling would lead to mindless exercises or crank-turning without physical understanding is rarely stated nowadays. Mindless calculations can be done without a computer and most researchers feel that the use of modern computational equipment allows boarder exploration of physical phenomena by reducing some of the constraints arising from mathematical and physical complexity. As a result we are in the fortunate situation for many subareas of condensed matter physics where theorists can contribute more effectively in the three conventional modes expected of them. Firstly, there is the task of explaining specific observed phenomena. Secondly, the development of concepts, arguments, or robust theories to bring together the results of a variety of experiments with unifying principles. Finally, there is prediction. The use of theory to predict new results is particularly challenging in the field of condensed matter physics where few predictions are ignored by able experimentalists. In practice it is not unusual to find theorists working in two or three modes at once.

The evolution of the theory of bonding and structural properties of solids is a particularly good example of the metamorphosis which has taken place more generally in condensed matter theory. It is also an area where successful predictions are being made. After a brief description of some of the approaches and models used, a few applications will be described as examples to illustrate the successes of the models and concepts and to indicate some future directions of this field.

2. Models

Because of the successes in the application of quantum theory to understand the properties of atomic and molecular systems, it was natural to view clusters of atoms and solids as weakly perturbed collections of atoms or molecules. This approach has been used successfully by quantum chemists and physicists for systems in which the bonding between the atoms or molecules does not significantly perturb their electronic energy levels. For most metals and semiconductors, the opposite is true. In fact, it is usually a better starting point is to consider free electrons weakly perturbed by the ionic potential. The unperturbed free electron model was used by Sommerfeld and others to explain many properties of metals (Kittel 1986). When this model is used, the Fermi statistics and spatial constraints on the electron gas manifest themselves in most of the physical properties particularly at low temperatures. To go beyond the simplest free-electron model which considers only the kinetic energy contribution, it is convenient to use a jellium model where the electrons move freely in a structureless positive jelly arising from a smeared out distribution representing the ions. The total electronic energy E_e (in rydbergs) of the system is composed of kinetic, exchange, and correlation energies, and all are functions of the electron gas parameter r_s,

$$E_e = 2.2099/r_s^2 - 0.9163/r_s - 0.094 + 0.0622 \ln r_s. \tag{1}$$

The above expression is accurate as $r_s < 0$ and for large solids where surface effects can be neglected. Even though $2 < r_s < 5$ for most metals, this is still a good starting point for bulk solids and with some refinements surface effects can be explored for finite samples.

A recent success of this model is its application to metal clusters. Jellium spheres and ellipsoids have been used to stimulate clusters of alkali atoms (de Heer *et al.* 1987). The confinement of the electrons in the jellium spheres and ellipsoids leads to electronic energy levels and shell structure which in turn are responsible for the stability of clusters with specific or 'magic' numbers of atoms. The self-consistent potential for the electrons resembles that of the intermediate well in figure 1 which gives magic numbers at 2, 8, 18, 20, 43, 40, ... This electronic shell structure is basic to the theory and the resulting calculations of the abundance spectra, ionization potentials and many other electronic properties of metal clusters are generally in agreement with the measured results. Because the jellium model is used, the structural arrangement of the atoms does not play a role in this approach. Calculations (Cleland & Cohen 1985) based on pseudopotentials for specific structural arrangements of atoms yield results which are consistent with the jellium spectra. Hence the crystal field effects do not significantly alter the shell structure.

When large clusters are formed (Göhlich *et al.* 1990) the transition to the bulk can be studied. For free-electron-like metals, small bonding features in the electron density appear when the cluster size is sufficiently large. A nearly-free-electron model with the electron–ion potential treated in perturbation theory can be used to describe this behaviour. A similar approach can be applied to semiconductors and even some insulators. The electron–ion potential can be approximated using a pseudopotential approach (Heine 1970; Cohen & Heine 1970) where the core electrons are assumed to

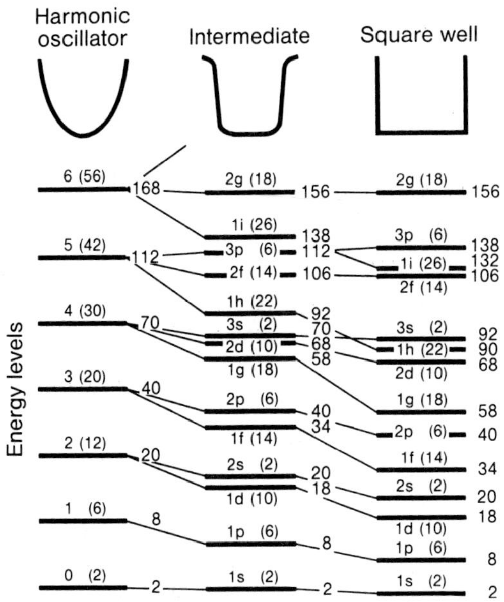

Figure 1. A comparison of the energy level spectrum for a three-dimensional harmonic oscillator (left), a square-well potential (right), and a potential intermediate between these two. The numbers inside the rectangles refer to energy level labels while those in parenthesis refer to degeneracies. The numbers outside the rectangles give the cumulative number of states. The intermediate case is representative of that found for simple metal clusters.

be unchanged in going from the nearly isolated atoms state appropriate for a gas of atoms to the strongly interacting atoms state of a solid. Now the total energy of the system includes the electronic energy, the energy of the ionic cores and the electron–core interaction energy. For rigid cores, the total energy is

$$E_{\text{total}} = E_{\text{ke}} + E_{\text{e-c}} + E_{\text{e-e}} + E_{\text{c-c}}, \tag{2}$$

where E_{ke} is the electron kinetic energy; $E_{\text{e-c}}$, $E_{\text{e-e}}$, and $E_{\text{c-c}}$ are the electron–core, electron–electron and core–core contributions to the energy.

The pseudopotentials needed to represent the electron–core interactions can be extracted from experimental data (Cohen & Chelikowsky 1988) or generated (Starkloff & Joannopoulos 1977; Zunger & Cohen 1979; Hamann *et al.* 1979; Kerker 1980; Louie *et al.* 1982) from atomic wavefunctions. The former approach is the empirical pseudopotential method (EPM) while the latter is often called the '*ab initio* or first principles' pseudopotential approach. The EPM generally starts with optical data and yields an interpretation of the data together with pictorial representation of the electronic charge density, the band structure, dielectric functions, densities of states, etc. Calculations in the 1960s and 1970s using the EPM gave rise to predictions of band structures which were later shown to be correct using angular resolved photoemission. These studies also introduced electron density plots and successfully predicted the bonding distribution of the electrons.

Since the EPM uses potentials derived for the bulk solid it is not appropriate for calculations which involve the rearrangement of charge such as occurs at surfaces, near defects, or as a result of solid–solid structural phase transitions. For these cases the core pseudopotentials are used together with a model for the electron–electron

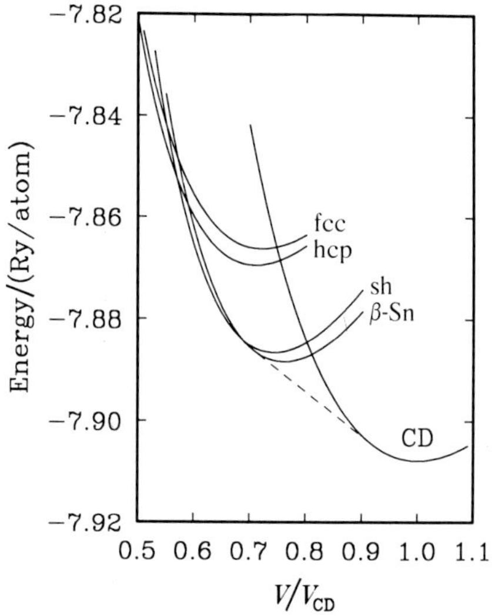

Figure 2. Calculated total energy as a function of volume for Si in the cubic diamond, β-Sn, simple hexagonal, hexagonal-close packed, and face-centred-cubic structures. The volume is normalized to the calculated volume for cubic diamond (V_{CD}).

interaction energy. A phenomenological approach to estimate the electron–electron potential can be used based on Slater's (1951)X–α model, or a more first principles scheme such as the local density approximation (LDA) (Kohn & Sham 1965) can be applied. The latter scheme is appropriate for calculating ground-state energies; it is particularly useful for determining structural and vibrational properties of solids. These properties can be computed (Cohen 1982) by considering the total energies of model systems for different structural arrangements. An example for Si is given in figure 2. This approach has had wide application for determining crystal structures at high pressures and mechanical properties such as compressibility. For vibrational properties, the atoms are moved to simulate a phonon distortion and then total energies or forces are computed to give phonon frequencies and related properties such as electron–phonon interactions.

Hence the use of pseudopotentials and the LDA allows broad studies of ground state properties of solids. Excited state properties such as those important for explaining optical processes require more complex schemes. A quasi-particle approach (Hybertsen & Louie 1986) for dealing with excited states relies on the calculation of the electron self-energy and local field effects. This method has been applied to several systems with success. Semiconductor and insulator band gaps which are underestimated using the LDA are found to be in good agreement with experiment when the quasi-particle spectrum is calculated using the above scheme. Furthermore, improvements for ground-state properties are found when electronic correlation effects are computed using quantum Monte-Carlo methods (Fahy *et al.* 1988).

Since the total energy approach usually relies on the comparison of energies for competing candidate structures or structural configurations (as shown in figure 2), it is limited by the choice of the candidates examined. Educated guesses and experimental data can be used to narrow the number of choices. If forces on the atoms in a specific structure are evaluated it is sometimes possible to determine the directions in which to move the atoms to obtain a minimum energy structure. This is particularly useful for surface structures where the number of possible distortions are more limited. A more promising approach (Car & Parrinello 1985) is to use Monte-Carlo sampling techniques for atoms moving randomly. Energy minimization procedures based on molecular dynamics schemes are developing at a rapid pace and it should be possible to explore a wide range of parameter space in the future.

3. Applications

As discussed previously, the EPM has a long list of applications (Cohen & Chelikowsky 1988) to electronic structure and optical properties. Its accuracy is rivalled only by the first-principles quasi-particle calculations (Hybertsen & Louie 1986). Since the EPM is much simpler it has had broader applications, but it requires experimental input. For ground-state properties, the total energy pseudopotential–LDA approach (Cohen 1982) has been applied most extensively. One impressive result is the successful prediction of hexagonal phases of Si, their structural properties, and the existence of superconductivity in these covalent metals systems (Chang *et al.* 1985). Other high-pressure phases have been predicted and their properties explained. It is gratifying that this first-principles approach which uses only the atomic number and atomic mass as input is capable of giving ground-state structural properties such as lattice constants, bulk moduli, the Poisson ratio, etc., with high precision. In addition vibration properties such as phonon spectra, Gruneisen constants, and the characteristic properties of solid-state structural transitions such as transition pressures and transition volumes are computed accurately.

Below two recent applications are discussed in some detail. These are the search for low compressibility or very hard solids and the properties of solid hydrogen at high pressures.

(*a*) *Low compressibility solids*

At present the largest bulk modulus and hardest known substance is diamond. Although hardness can be influenced by macroscopic defects, for ideal crystals, the bulk modulus is the best indicator of hardness. Since the bulk modulus depends on the microscopic bonding of the solid, it can be calculated from first principles. One direct approach for calculating the bulk modulus is by using the total pseudopotential method to compute the dependence of the energy on volume. Using a Murnahan (1944) or Birch (1952) equation of state to fit the calculated curves, the bulk modulus and its pressure derivative can be obtained. Although this approach works well, it requires considerable computer time. A simpler but empirical scaling approach is described below. This method allows the investigation of material trends and provides simple formulae for estimating bulk moduli.

The scaling approach (Cohen 1985) is based on the following observations. For a free electron system the bulk modulus

$$B = \tfrac{2}{3} n E_{\mathrm{F}}, \tag{3}$$

where n is the electron concentration and E_{F} is the Fermi energy. Equation (3) can be interpreted as a scaling relation for B in which the bulk modulus scales like the bonding energy divided by the bond volume. The extension to covalent semiconductors is based on the spectral model developed by Phillips and Van Vechten

(Phillips 1984). In this model the average optical gap E_g is assumed to have a homopolar contribution E_h and an ionic contribution C, where

$$E_g^2 = E_h^2 + C^2. \tag{4}$$

The values of E_g, E_h and C can be extracted from experiment and their dependence on bond length can be determined (Phillips 1984).

In a series such as the Ge, GaAs, ZnSe row of the periodic table, E_h and the lattice constant remain unchanged. It is the ionic contribution C which changes the average gap which appears in the optical spectrum. The covalent nature of the bond determines the bond length and is the major factor in determining the size of the bulk modulus. If the bond is taken to be cylindrical with a radius of order the Bohr radius and a length d, then the scaling of B discussed in relation to equation (3) suggests that

$$B = 45.6\, E_h\, d^{-1}, \tag{5}$$

where E_h (in electronvolts) is considered to be the bond energy and B is given in GPa when d is in ångströms†. Since E_h scales as $d^{-2.5}$, a relation which depends only on d can be derived

$$B = 1761\, d^{-3.5}. \tag{6}$$

In the above discussion the ionic factor is ignored. A simple extension to include ionicity can be made by introducing a parameter $\lambda = 0, 1, 2$ for group IV, III–V and II–VI materials. The resulting equation

$$B = \tfrac{1}{4}N_c\,(1971 - 220\,\lambda)\, d^{-3.5} \tag{7}$$

gives agreement within a few percent of the measured values of the bulk modulus of valence 8 compounds where $N_c = 4$. A generalization to non-octet compounds which do not have complete tetrahedral bonding can be made if one considers the density of the bonds in addition to the strength of the bonds. Defining N_c as the average coordination number in the crystal, then the factor $\tfrac{1}{4}N_c$ multiplies the result in equation (7).

As is evident from (7), the bulk modulus increases as d and λ decrease. Although diamond sets the limit at present there is no *a priori* reason to expect that the value $B = 443$ GPa for diamond is an upper limit. In particular it was suggested (Cohen 1985) that compounds based on carbon and nitrogen might exhibit values of B comparable with that of diamond. The C–N bond length is short and this bond is not very ionic. The empirical scaling approach does not predict d, however, if atomic radii are used to estimate d, the expected values of B are very large, Unfortunately, complete tetrahedral coordination is not expected since a zincblende structure with an occupied antibonding band does not appear favourable.

To test the viability of having a C–N compound with a very high B, the first-principles total energy approach was used (Liu & Cohen 1989). This method requires a candidate crystal structure. The structure chosen is the β-Si_3N_4 structure (figure 3) which is known to exist. Assuming this structure, the ground-state properties of β-C_3N_4 are computed using the pseudopotential total-energy approach within a localized-orbital formalism (Chelikowsky & Louie 1984). The hexagonal unit cell chosen contains two formula units and the structure can be viewed as a C–N network with sp^3 hybrids on the C atoms and sp^2 hybrids on the N atoms. This three-dimensional network is rigid and is a good prototype for achieving a low compressible form of C–N even though the bonding is not completely sp^3.

† 1 Å = 10^{-10} m.

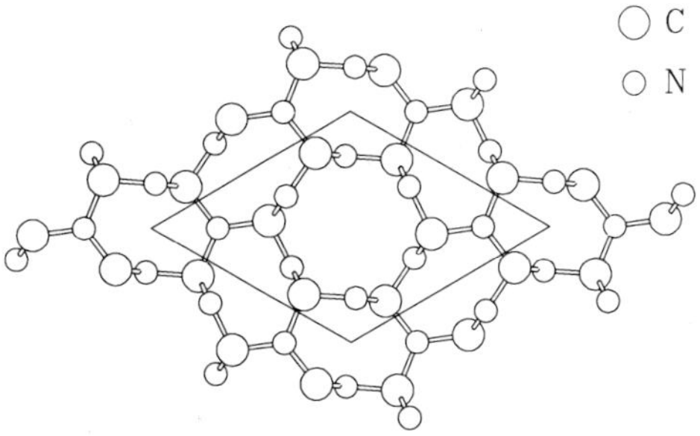

Figure 3. Structure of β-C_3N_4 in the *a–b* plane. The *c*-axis is normal to the page. Half of the atoms illustrated are located in the $z = -\frac{1}{4}c$ plane and the other half are in the $z = \frac{1}{4}c$ plane. The parallelogram indicates the unit cell.

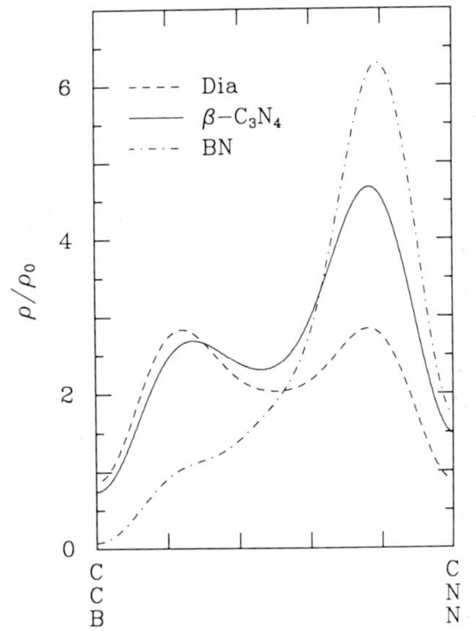

Figure 4. Comparison of the normalized valence charge density along the bond for diamond, β-C_3N_4 and BN. The normalization factor for the charge density, ρ_0, is the number of valence electrons per cell.

The calculated total energy against volume for β-C_3N_4 yields a C–N bond length of 1.47Å which is intermediate between the sum of the C tetrahedral and N sp^2 radii and the sum of C and N tetrahedral radii. At the equilibrium volume, the cohesive energy is 81 eV per cell or an average of 5.8 eV per atom which suggests that this structure has a reasonable chance of being metastable. The bulk modulus is calculated to be 427 ± 15 GPa which is comparable to the value of 443 GPa for diamond. Using the calculated mass density and bulk modulus, the average bulk velocity of sound is estimated to be 1.1×10^6 cm s^{-1}. The electronic charge density along the C–N bond is similar to that of a C–C bond in diamond except for some excess charge on the N site. A comparison with diamond and BN is given in figure 4.

The prototype β-C_3N_4 calculation demonstrates that solids based on C and N can be metastable and if formed should have very large bulk moduli. In addition, the good agreement between the estimates of the bulk modulus from equation (7) and from the total energy calculation adds support for the use of the scaling formula to investigate trends. It is hoped that experiments using high temperatures, high pressures, molecular beam epitaxy or plasma deposition will lead to the fabrication of β-C_3N_4 or similar phases. Although detailed calculations were not done for α-C_3N_4 (C–N in the α-Si_3N_4) structure, this structure should also be a favourable candidate. Amorphous phases of C_3N_4 are also of considerable interest. In addition to their low compressibility, the wide band gaps and high thermal conductivities expected for C–N compound make these materials interesting from a technological as well as a scientific point of view.

(*b*) *Metallic hydrogen*

Although solid hydrogen is an insulator at low temperatures and pressures, there is great interest in exploring higher-pressure solid phases which are likely to be metallic. Wigner & Huntington (1935) suggested that at high pressures the H_2 molecular bond would be broken and a monatomic metallic system would result. Another possibility for a metallization is an insulator–metal transition caused by band overlap. The resulting molecular metal would still have strong short H_2 bonds with longer bonds linking the dimers. Theoretical estimates of the pressure for the transition to an atomic metal are in the range of 3–5 mbar† (Min *et al.* 1986; Ceperly & Alder 1987; Barbee *et al.* 1989*a*). Estimates of the transition pressure to a molecular metal phase are at much lower values in the range of 2 mbar (Min *et al.* 1986; Barbee *et al.* 1989*a*; Ramaker *et al.* 1975; Friedli & Ashcroft 1977). These calculations are not definitive because the structural phases of the high-pressure insulator and metallic phases are not known.

Recently theoretical activity in this area has intensified because of the interpretation of some experimental data as evidence of metallization (Mao & Hemley 1989; Eggert *et al.* 1990). Several experimental studies in the range of 150 GPA suggest there is a transition to a new phase, hydrogen-A (H-A). The measurements indicate that below the transition the structure is hcp and continuity of the Raman active phonon mode suggests that the H-A phase is also hcp, but there is a discontinuous change in the frequency of the vibron mode near the transition. Some optical data (Mao *et al.* 1990) suggest free carriers and Drude behaviour.

To investigate the possibility of band overlap around 150 GPa, an LDA calculation was modified using an X–α potential (Garcia *et al.* 1990). The band structure calculation is done for an oriented arrangement of the molecular axes and for a disordered phase modelled by a structure factor which produces spherically symmetric charge distributions. Using experimental input for structural properties and the zero pressure band gap, the density dependence of the minimum indirect gap is determined for the oriented and spherically symmetric phases. As shown in figure 5 band overlap occurs at a much higher density for the latter phase. When the experimental equation of state is used, the calculated critical densities lead to estimates of 180 ± 20 GPa and 400 ± 40 GPa for the transition pressures giving band overlap for the oriented and spherically symmetric phases, respectively.

Although the LDA calculations significantly underestimate the transition pressures as expected, the ordering found in the X–α calculation is the same. Previous comparisons between LDA, X–α, and quasi-particle calculations indicate that LDA estimates are low while X–α estimates are a bit high compared with quasi-particle calculations for the density corresponding to a zero gap transition. Recently (Chacham & Louie 1991) have completed a first-principles quasi-particle study of the oriented phase and estimate that overlap occurs at 150 GPa.

† 1 bar = 10^5 Pa.

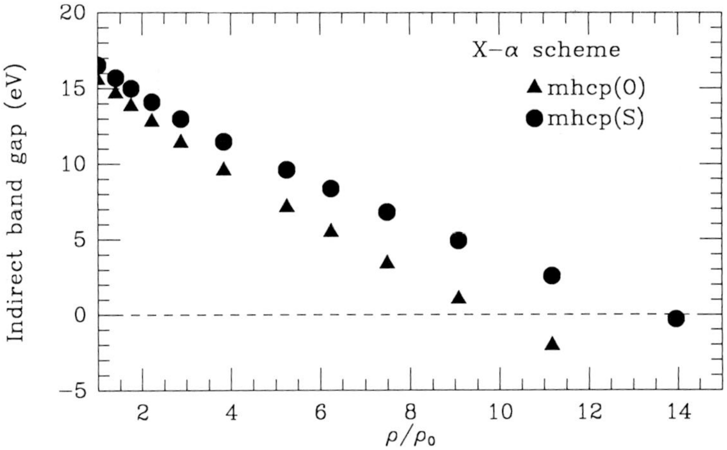

Figure 5. Indirect band gap against density for molecular-hexagonal-close-packed hydrogen assuming oriented molecules (triangles) and spherically symmetric molecules (solid circles). The calculation was done using the Slater X–α exchange potential.

The transition to the monatomic metallic phase can be studied by examining a variety of structural candidates and comparing their enthalpies as a function of pressure. It was shown (Barbee *et al.* 1989*a*) that the primitive hexagonal (PH) phase is favoured relative to several other common structures. However, when the phonon spectra was calculated (Barbee *et al.* 1989*b*) imaginary harmonic phonon frequencies were found indicating that a transition to another structure is favoured. By tripling the unit cell along the hexagonal *c* axis a lowering of the energy was found. Although this distorted hexagonal structure has a low energy, some phonon modes appear to be unstable signalling yet another transition. An interesting low-energy structure is the 9*R* structure found in Li at low temperatures; calculations based on this structure are in progress.

By calculating the phonon spectra for various low-energy structures and electron–phonon matrix elements, estimates of the electron–phonon parameter $\lambda \approx 1$ are made. Although this is a moderately large value of λ, the presence of high-energy phonons arising from the low hydrogen mass leads to estimates of superconducting transition temperatures T_c in the range of 100–200 K. The value of T_c depends on crystal structure and the resulting electronic properties. Because of similarly expected high phonon frequencies in the molecular metal, if the electron–phonon matrix elements are comparable with those found for the monatomic metal, then T_c should also be large for this system.

4. Conclusions

The emphasis here is on the robustness of the pseudopotential total energy approach. Its evolution from the nearly-free-electron gas model and the development of refinements to the model have provided a useful tool and significant physical insight into the behaviour of solids and clusters. There have been dozens of applications; only a few are highlighted here. Possible future paths including taking advantage of the developments in calculating excited states, correlation energies, and molecular dynamics schemes. Once these approaches are refined and simplified, it should be possible to use one computational approach to compute ground-state and excited-state properties for previously observed or new hypothetical systems obtained by Monte-Carlo type searches.

If the above goals are achieved it is likely that they will lead to even more ambitious undertakings. However, regardless of future advances, it is encouraging to note that even now successful predictions are being made of new materials and new properties. And as described earlier, when the methods are pushed to their present limits, there are the suggestions of materials with hardness comparable with diamond and a superconductor with a transition temperature above any found thus far.

This work was supported by NSF Grant no. DMR88-18404 and by the Director, Office of Energy Research, Office of Basic Energy Sciences, Materials Science Division of the U.S. Department of Energy under Contract no. DE-AC03-76SF00098.

References

Barbee, III, T. W., Garcia, A. & Cohen, M. L. 1989*b* First principles prediction of high temperature superconductivity in metallic hydrogen. *Nature, Lond.* **340**, 369–371.

Barbee, III, T. W., Garcia, A., Cohen, M. L. & Martins, J. L. 1989*a* Theory of high-pressure phases of hydrogen. *Phys. Rev. Lett.* **62**, 1150–1153.

Birch, F. 1952 Elasticity and constitution of the Earth's interior. *J. geophys. Res.* **57**, 227–286.

Car, R. & Parrinello, M. 1985 Unified approach for molecular dynamics and density-functional theory. *Phys. Rev. Lett.* **55**, 2471–2474.

Ceperley, D. M. & Alder, B. J. 1987 Ground state of solid hydrogen at high pressures. *Phys. Rev.* B **36**, 2092–2106.

Chacham, H. & Louie, S. G. 1991 Metallization of solid hydrogen at megabar pressures: a first-principles quasiparticle study. *Phys. Rev. Lett.* **66**, 64–67.

Chang, K. J., Dacorogna, M. M., Cohen, M. L., Mignot, J. M., Chouteau, R. & Martinez, G. 1985 Superconductivity in high-pressure metallic phases of Si. *Phys. Rev. Lett.* **54**, 2375–2378.

Chelikowsky, J. R. & Louie, S. G. 1984 First-principles linear combination of atomic orbitals method for the cohesive and structural properties of solids: application to diamond. *Phys. Rev.* B **29**, 3470–3481.

Cleland, A. N. & Cohen, M. L. 1985 Pseudopotential calculation of shell structure in sodium clusters. *Solid St. Commun.* **55**, 35–38.

Cohen, M. L. & Chelikowsky, J. R. 1988 *Electronic structure and optical properties of semiconductors.* (264 pages.) Berlin: Springer-Verlag.

Cohen, M. L. & Heine, V. 1970 The fitting of pseudopotentials to experimental data and their subsequent application. *Solid state physics 24* (ed. H. Ehrenreich & D. Turnbull), pp. 37–248.

Cohen, M. L. 1982 Pseudopotentials and total energy calculations. *Physica Scr.* **T1**, 5–10.

Cohen, M. L. 1985 Calculation of bulk moduli of diamond and zinc-blende solids. *Phys. Rev.* B **32**, 7988–7991.

Cohen, M. L. 1986 Predicting new solids and superconductors. *Science, Wash.* **234**, 549–553.

de Heer, W. A., Knight, W. D., Chou, M. Y. & Cohen, M. L. 1987 Electronic shell structure and metal clusters. *Solid St. phys. 40* (ed. H. Ehrenreich & D. Turnbull), pp. 93–181.

Eggert, J. H., Goettel, K. A. & Silvera, I. F. 1990 High-pressure dielectric catastrophe and the possibility that the hydrogen-A phase is metallic. *Europhys. Lett.* **11**, 775–781.

Fahy, S., Wang, X. W. & Louie, S. R. 1988 Variational quantum Monte Carlo nonlocal pseudopotential approach to solids: cohesive and structure properties of diamond. *Phys. Rev. Lett.* **61**, 1631–1634.

Friedli, C. & Ashcroft, N. W. 1977 Combined representation method for use in band-structure calculations: application to highly compressed hydrogen. *Phys. Rev.* B **16**, 662–672.

Garcia, A., Barbee, III, T. W., Cohen, M. L. & Silvera, I. F. 1990*a* Band gap closure and metallization of molecular solid hydrogen. *Europhys. Lett.* **13**, 355–360.

Göhlich, H., Lange, T., Bergmann, T. & Martin, T. P. 1990 Electronic shell structure in large metallic clusters. *Phys. Rev. Lett.* **65**, 748–751.

Hamann, D. R., Schlüter, M. & Chiang, C. 1979 Norm-conserving pseudopotentials. *Phys. Rev. Lett.* **43**, 1494–1497.

Heine, V. 1970 The pseudopotential concept. *Solid state physics 24* (ed. H. Ehrenreich & D. Turnbull), pp. 1–36.

Hybertsen, M. S. & Louie, S. G. 1986 Electron correlation in semiconductors and insulators: band gaps and quasiparticle energies. *Phys. Rev.* B **34**, 5390–5413.

Kerker, G. P. 1980 Non-singular atomic pseudopotentials for solid state applications. *J. Phys.* C **13**, L189–L194.

Kittel, C. 1986 *Introduction to solid state physics*, 6th edn. (646 pages.) New York: Wiley.

Kohn, W. & Sham, L. J. 1965 Self-consistent equations including exchange and correlation effects. *Phys. Rev.* **140**, A1113–A1118.

Liu, A. Y. & Cohen, M. L. 1989 Prediction of new low compressibility solids. *Science, Wash.* **245**, 841–842.

Louie, S. G., Froyen, S. & Cohen, M. L. 1982 Nonlinear ionic pseudopotentials in spin-density-functional calculations. *Phys. Rev.* B **26**, 1738–1742.

Mao, H. K. & Hemley, R. J. 1989 Optical studies of hydrogen above 200 gigapascals: evidence for metallization by band overlap. *Science, Wash.* **244**, 1462–1465.

Mao, H. K., Hemley, R. J. & Hanfland, M. 1990 Infrared reflectance measurements of the insulator–metal transition in solid hydrogen. *Phys. Rev. Lett.* **65**, 484–487.

Min, B. I., Jansen, J. F. & Freeman, A. J. 1986 Pressure-induced electronic and structural phase transitions in solid hydrogen. *Phys. Rev.* B **33**, 6383–6390.

Murnaghan, F. D. 1944 The compressibility of media under extreme pressures. *Proc. natn. Acad. Sci. U.S.A.* **30**, 244–247.

Phillips, J. C. 1984 *Bonds and bands in semiconductors.* (288 pages.) New York: Academic.

Ramaker, D. E., Kumar, L. & Harris, F. E. 1975 Exact-exchange crystal Hartree–Foch calculations of molecular and metallic hydrogen and their transitions. *Phys. Rev. Lett.* **34**, 812–814.

Slater, J. C. 1951 A simplification of the Hartree–Fock Method. *Phys. Rev.* **81**, 385–390.

Starkloff, T. & Joannopoulos, J. D. 1977 Local pseudopotential theory for transition metals. *Phys. Rev.* B **16**, 5212-5215.

Wigner, E. & Huntington, H. B. 1935 On the possibility of a metallic modification of hydrogen. *J. chem. Phys.* **3**, 764–770.

Zunger, A. & Cohen, M. L. 1979 First principles nonlocal-pseudopotential approach in the density functional formalism. II. Application to electronic and structural properties of solids. *Phys. Rev.* B **20**, 4082–4108.

Discussion

Z. Reut (*City University, London, U.K.*). The discriminating ability of the pseudo-potential should be quantified; i.e. what are changes in the pseudo-potential when going from lighter to heavier elements, up and down in the Mendeleev periodic system of elements? This is related to the effectiveness of the predictive methods.

M. L. Cohen. There are some standard variations which are quite interesting; these are well described by the pseudopotential approach. As an example, let me choose column IV of the periodic table. For carbon, there are no p states in the core and as a consequence no repulsive p core pseudopotential. The valance p electrons lie close to their atomic spatial positions in C whereas for Si they are pushed out toward the centre of the Si–Si bond. The absence of d core states for Si has similar effects on the d valence electrons and the metallic component of the bonding which increases as one moves to Sn and Pb. Upon reaching the Pb row, relativistic effects become more important and the s and p states move apart in energy making it more difficult to construct sp^3 orbitals. The pseudopotential calculations of material properties illustrate these trends. For example the pressure-induced structural changes of solids are influenced by the core properties discussed above. These are evident in pseudopotential calculations of the transition pressures for solid–solid phase transformations.

L. J. SHAM (*University of California, San Diego, U.S.A.*). For calculation of the superconducting transition temperature in hydrogen, what does Professor Cohen use for electron repulsion? Hydrogen appears to be exotic compared with the materials McMillan considered for μ^*.

M. L. COHEN. Our level of ignorance regarding the electron–electron repulsion parameter for hydrogen is the same as for other metals. We use the standard procedure of scaling μ with r_s and then convert to a μ^* model. In the end the value for atomic metallic hydrogen is not unusual. In this regard, hydrogen is not terribly exotic.

P. B. ALLEN (*SUNY, Stony Brook, U.S.A.*). BCC–Li would probably be superconducting, but at low T lithium becomes 9R which seems to kill superconductivity. Would this also happen in hydrogen? Does Professor Cohen understand why this happens in Li?

M. L. COHEN. We didn't calculate the electron–phonon coupling constants and superconducting transition temperature for BCC hydrogen because this is an unstable structure, however, I wouldn't be surprised if λ and T_c were larger for BCC than for $9R$ hydrogen. We are examining Li now.

N. W. ASHCROFT (*Cornell University, U.S.A.*). There are some quite simple systems (aluminium is an example) that possess moderately high bulk moduli, but otherwise would be characterized as 'weak', i.e. the shear strengths are low. It seems that a real test of 'superhardness' would require both stability analysis of a proposed structure (against, for instance, shear distortion) and if stable a determination of the shear modulus on shear strength. Given this, can Professor Cohen report on the shear strength of the carbon–nitride system, and of the other systems that are being referred to as superhard?

M. L. COHEN. The microscopic property which is the best indicator of hardness is the bulk modulus. However, macroscopic properties such as defects and dislocations effect the hardness of real materials, and as Professor Ashcroft points out the shear strength of a material can be important in this regard. For the covalent materials I discussed this is less of a problem, and for BN and B–C_3N_4 our calculations suggest fairly isotropic properties.

V. HEINE (*Cambridge, U.K.*). (*a*) As regards sensitive quantities like band offsets, there have been a few unsatisfactory calculated results when some atoms like Zn and Cd were involved. Work this year by Qteish and Needs in Cambridge has put into the calculations the so-called nonlinear core corrections to the LDA treatment of exchange and correlation. When this is done, much better agreement with experiment is obtained, thus adding to the success story that Professor Cohen has outlined. Of course they are now going on to more complicated situations that are not clear from experiment.

(*b*) In answer to the question that was asked about a pseudopotential for hydrogen: Richard Needs in published calculations with Richard Martin found that the use of a pure Coulomb potential for hydrogen gave better convergent results for energy differences between structures than the use of a hydrogen pseudopotential.

(*c*) Jürgen Hafner and I obtained some interesting indications about the structure of metallic hydrogen using perturbation theory which we never had time to follow up. Now I know perturbation theory is rather ignored in these days of total energy calculations. Moreover the pairwise interactions from perturbation theory expressed in real space has been under a cloud because calculations by different workers in the 1960s and early 1970s gave significantly different results. The latter point has been largely resolved, particularly by Hafner (Hafner & Heine 1986). It is necessary to use a good formula for the exchange and correlation contribution to the dielectric function which satisfies the compressibility sum role. This controls sensitively the radius of the hard core repulsive interaction and hence the shape of the pair potential in the crucial region where the screened core repulsion goes over into the Friedel oscillations with their phase shift involving the pseudopotential. When that is done correctly, one gets pair potentials which yield a great deal of insight into the structures of the elements, even the covalently bonded structures (Hafner & Heine 1983). Anyway, when one calculates such a pair potential for hydrogen in the metallic range of $r_s = 1.2$–1.7 one finds bumps relative to the close packed distance similar to those in Ga, In, Hg which give strange structures there. This is fully in accordance with Professor Cohen's experience that simple structures are unstable. But I found it difficult to come to any clear conclusion about what the coordination number should be and I wonder whether Professor Cohen's search for the correct ground-state structure is at an end.

M. L. Cohen. I agree with Professor Heine's comments. For (*a*), we have had success in calculating the quasi-particle band offset at the (001) interface for GaAs–AlAs heterojunctions (Zhang *et al.* 1990). For many problems it is necessary to use the procedures exchange and correlation interactions. Professor Heine's examples involving Zn and Cd are good ones. Even structural properties are affected by these nonlinear core effects. (*b*) I agree that for most applications it is best to use the full Coulomb potential. (*c*) Kogan and others, in the U.S.S.R. also claim success when using perturbation theory. I am glad to hear that Professor Heine's results for the stability of the simple structures are similar to ours. I do not know whether we have found *the* ground-state structure. At this point the 9*R* structure appears to be stable, and it has a low energy. Perhaps a lower-energy structure will turn up in the future.

Additional references

Hafner, J. & Heine, V. 1983 *J. Phys.* F **13**, 2479–2501.

Hafner, J. & Heine, V. 1986 *J. Phys.* F **16**, 1429–1458.

Zhang, S. B., Cohen, M. L., Louie, S. G., Tomanék, D. & Hybertsen, M. S. 1990 Quasiparticle band offset at the (001) interface and band gaps in ultrathin superlattice of GaAs–AlAs heterojunctions. *Phys. Rev.* B **41**, 10058–10067.

Order and disorder in metallic alloys

By B. L. Gyorffy[1], G. M. Stocks[2], B. Ginatempo[3], D. D. Johnson[4], D. M. Nicholson[2], F. J. Pinski[5], J. B. Staunton[6] and H. Winter[7]

[1] *H. H. Wills Physics Laboratory, University of Bristol, Tyndall Avenue, Bristol BS8 1TL, U.K.*

[2] *Metals and Ceramics Division, Oak Ridge National Laboratory, P.O. Box 2008, Oak Ridge, Tennessee 37831-6114, U.S.A.*

[3] *Instituto di Fisica Teorica, Universitá di Messina, Messina, Italy*

[4] *Sandia National Laboratories, P.O. Box 969, Livermore, California 94551-0969, U.S.A.*

[5] *Department of Physics, University of Cincinnati, Cincinnati, Ohio 45221, U.S.A.*

[6] *Department of Physics, University of Warwick, Coventry CV4 7AL, U.K.*

[7] *Kernforschungszentrum Karlsruhl, D-7500 Karlsruhl, F.R.G.*

Self-consistent 'band theory', based on density functional theory, is a useful approach to describing the electron glue which holds solids together. However, its powerful group theoretic and numerical techniques cannot be deployed for disordered states of matter. The self-consistent KKR-CPA is an analogous method which is able to deal with some of these interesting cases. In particular, we show how it describes random metallic alloys, treating all the classic Hume-Rothery factors: size-effect, electronegativity and electrons per atom ratio ($e:a$) on more or less equal footing and from first principles. Moreover, we use the KKR-CPA framework to analyse the instability of the disorder state to compositional ordering processes and hence provide a first principle description of the forces which drive order–disorder transformations.

1. Introduction

The electronic structure of random crystalline metallic alloys is an important chapter in the quantum theory of solids (Hume-Rothery & Coles 1969; Mott & Jones 1936). After atoms, molecules and ordered crystalline solids this is the next most tractable problem of positively charged nuclei and the 'electron glue' between them. Moreover, it is the simplest non-trivial example where the theme of order–disorder is fully developed. In short the electronic states have new features and they drive novel phenomena such as phase separation and compositional ordering. We review the current state of our first principles' understanding of the electronic structure and the way it determines the compositional short- and long-range order (de Fountain 1979; Kchaheturyan 1986).

Central to our discussion below are two assumptions: one is that of a rigid lattice and the other is that specifying the occupancy variable ξ_i (which takes on the value 1 if there is an A atom at the ith site and 0 if the atom at $\boldsymbol{R}_i$ is the B type) for all sites defines all compositional configurations. The first of these may be relaxed in favour of a symmetry statement, that on average the system is invariant under a set of translations which define a lattice, without seriously affecting the conceptual framework we are about to describe. The second, which is essentially the adiabatic theorem, can be relinquished only at the expense of an entirely new start involving diffusion. In short we consider configurations specified by the sets $\{\xi_i\}$ and average over all such sets with appropriate weights.

Traditionally, the study of electronic states in random alloys always followed that for pure metals and order intermetallic compounds with a lag, reflecting the technical difficulties of solving the Schrödinger equation without the full support of the Bloch theorem (Elliott *et al.* 1974; Ehrenreich & Schwartz 1976). One of the points we emphasize here is that recently the work on random alloys has caught up with that for ordered systems in the sense that we can now base our description of the electronic structure of each configuration $\{\xi_i\}$ on the local density approximation (LDA) to the density functional theory (DFT). Namely, for each configuration $\{\xi_i\}$, we consider the Kohn–Sham equation for the Greens function:

$$\left(\epsilon+\nabla^2-\sum_i v_\sigma^{\mathrm{LDA}}(\boldsymbol{r}-\boldsymbol{R}_i;\{\xi_i\})\right)G_{\sigma\sigma}(\boldsymbol{r},\boldsymbol{r};\epsilon)=\delta(\boldsymbol{r}-\boldsymbol{r}'), \tag{1}$$

where each potential well, centred at the atomic nuclei, is described by the LDA potential functional (Kohn & Vashista 1983), $v_{\mathrm{LDA}}^{\mathrm{eff}}(\boldsymbol{r}-\boldsymbol{R}_i;[n(\boldsymbol{r};\{\xi_i\}])$ and the charge density, $n(\boldsymbol{r};\{\xi_i\})$, is given by

$$n(\boldsymbol{r};\{\xi_i\})=-\frac{1}{\pi}\sum_\sigma\int \mathrm{d}\epsilon f(\epsilon)\,\mathrm{Im}\,G_{\sigma\sigma}(\boldsymbol{r},\boldsymbol{r};\epsilon). \tag{2}$$

Above, as usual, $f(\epsilon)$ stands for the Fermi function, $f(\epsilon)=(\exp\{\beta(\epsilon-\mu)\}+1)^{-1}$ at the inverse temperature $\beta=(k_{\mathrm{B}}T)^{-1}$ and electronic chemical potential μ, and σ is the spin label.

Thus, the starting point for our arguments is the following 'Gedanken' procedure: solve (1) and (2) self-consistently for a fixed configuration $\{\xi_i\}$, then, evaluate the DFT formula for the grand potential:

$$\Omega_e(\{\xi_i\})=\Omega_e[n(\boldsymbol{r};\{\xi_i\})] \tag{3}$$

and calculate the partition function for the combined electron nuclei system:

$$Z=\sum_{\{\xi_i\}}\exp\{-\beta(\Omega_e(\{\xi_i\})-\sum_i v_i\,\xi_i)\}, \tag{4}$$

where v_i is the local chemical potential difference $v_i=v_i^{\mathrm{A}}-v_i^{\mathrm{B}}$ for the two species of nuclei. As usual, v_i is allowed to vary from site to site for formal purposes only. Finally, the ensemble of configurations, $\{\xi_i\}$, is defined by the distribution function

$$P(\{\xi_i\})=Z^{-1}\exp\{-\beta(\Omega_e-\sum_i v_i\,\xi_i)\}. \tag{5}$$

2. A first principles mean-field theory of compositional order

Although the above programme is not tractable as it stands, its mean field theoretic version is. We demonstrate, by using non-trivial examples, that this is a dramatic step forward in understanding the basic physics of metallic alloys.

The mean field approximation to the theory specified by equations (1)–(4) has three parts (Kohn & Vashista 1983; Gyorffy & Stocks 1983; Gyorffy *et al.* 1989).

The first is the usual assumption of a form for the free energy (grand potential) Ω, as a function of the local concentration configurations:

$$\Omega(T, v; \{c_i\}) = \Omega_e(\{c_i\}) + k_B T \sum_i (c_i \ln c_i + (1-c_i) \ln (1-c_i)) - \sum_i v_i c_i, \tag{6}$$

where c_i is the thermal average of ξ_i, consistent with the form of (6) (e.g. $c_i = -\partial\Omega/\partial v_i$), and

$$\Omega_e(\{c_i\}) = \langle \Omega_e\{\xi_i\}) \rangle_0. \tag{7}$$

To be consistent with the simple form of the entropy contribution in (6) the average $\langle\ \rangle_0$ is to be taken with respect to the inhomogeneous product distribution

$$P_0(\{\xi_i\}) = \prod_i P_i(\xi_i) \tag{8}$$

for which each factor is parametrized by the local concentration c_i as follows:

$$P_i(\xi_i) = c_i \xi_i + (1-c_i)(1-\xi_i). \tag{9}$$

Finally, the state of compositional order is determined by finding the minimum of the grand potential, in (6), as a function of the local concentrations. Namely, at a given temperature T and chemical potential difference $v(= v_i \forall i)$ the equilibrium concentration configuration $\{\bar{c}_i\}$ is the solution of the Euler–Langrange equation:

$$(\partial\Omega/\partial c_i)_{\{\bar{c}_i\}} = 0. \tag{10}$$

The second and third part of the statement, which constitutes the mean field theory, has to do with the calculation of the average electronic grand potential specified by equations (7)–(9).

In the spirit of the LDA and the mean field theory the second part simplifies the local potential function in (1) by replacing it by its local, partial average,

$$\bar{v}(\boldsymbol{r}-\boldsymbol{R}_i; \xi_i) = \xi_i v^{A}(\boldsymbol{r}-\boldsymbol{R}_i; [\bar{n}^{A}(\boldsymbol{r}), \bar{n}(\boldsymbol{r})]) + (1-\xi_i) v^{B}(\boldsymbol{r}-\boldsymbol{R}_i; [\bar{n}^{B}(\boldsymbol{r}), \bar{n}(\boldsymbol{r})]), \tag{11}$$

where $\bar{n}^{A}(\boldsymbol{r})$ and $\bar{n}^{B}(\boldsymbol{r})$ are the partly averaged charge densities, $\bar{n}(\boldsymbol{r})$ is the fully averaged charge density:

$$\bar{n}(\boldsymbol{r}) = c\bar{n}^{A}(\boldsymbol{r}) + (1-c)\bar{n}^{B}(\boldsymbol{r}) \tag{12}$$

and the potential functionals $v^{\alpha}(r-R_i; [\bar{n}^{\alpha}(\boldsymbol{r}), \bar{n}(\boldsymbol{r})])$ are the usual LDA functional using $\bar{n}^{\alpha}(\boldsymbol{r})$ for the contributions from the ith unit cell and $\bar{n}(\boldsymbol{r})$ for the contributions from all the other unit cells. Evidently, α is A or B. Note that for the sake of clarity we have assumed that we are working in the disordered state where all A sites are equivalent and hence all A sites are characterized by the same partly averaged charge density $\bar{n}^{\alpha}(\boldsymbol{r})$, and potential function $\bar{v}^{A}(r-R_i; \bar{n}^{A}(\boldsymbol{r}), \bar{n}(\boldsymbol{r}))$. Clearly, the equivalent statement applies for the B sites.

The final, and third, part splits into two logically separate instructions: the first is to solve

$$\left(\epsilon + \nabla^2 + \sum_i \bar{v}(r-R_i; \xi_i)\right) G(\boldsymbol{r}, \boldsymbol{r}; \epsilon) = \delta(\boldsymbol{r}-\boldsymbol{r}') \tag{13}$$

for the averaged and partly averaged Greens functions and charge densities

$$\bar{n}^{\alpha}(\boldsymbol{r}) = -\frac{1}{\pi}\int d\epsilon f(\epsilon)\, \mathrm{Im}\, \langle G(\boldsymbol{r}, \boldsymbol{r}; \epsilon)\rangle_0^{i\alpha}, \tag{14}$$

calculate the new potentials using (11), and repeat the procedure until convergence. The second is to use the coherent potential approximation (CPA) (Elliott *et al.* 1974; Ehrenreich & Schwartz 1976) for calculating the averages and hence take for $\Omega_e(\{c_i\})$ in (7) $\Omega_e^{\mathrm{CPA}}(\{c_i\})$ (Gyorffy *et al.* 1989).

Note that the first of these instructions interchanges the order of the self-consistency procedure demanded by DFT and the statistical averaging over the ensemble of configurations. Evidently, this is one of the features of the scheme which renders it tractable. The second instruction is also worthy of comment. Clearly the CPA is the natural approximation to use because, as it is made explicit by (8), the occupation variables are statistically independent and under such circumstances the CPA is known to be the mean field theory of disorder (Schwartz & Sigga 1972). Moreover, the self-consistent CPA algorithm which is implied by the above procedures have been fully implemented for realistic, muffin-tin, crystal potentials by the SCF-KKR-CPA method (Stocks & Winter 1984). This is the principle computational advance which makes the proposed calculations a practical proposition.

To highlight the physical content of the above first-principles mean field theory, we note that the solution of (10) in the disordered state is $\bar{c}_i = \bar{c}\,\forall i$. If the solution takes on any other pattern we speak of long-range order. Regions of the v, T plane, where solutions of different symmetry are of the lowest free energy, are separated by lines of phase boundaries and the totality of these constitutes the alloy phase diagram. Short-range order, on the other hand, is described by various derivatives of Ω with respect to the chemical potential. For example the Warner–Cowley short-range order parameter α_{ij} is given by the relation

$$\alpha_{ij} = \frac{1}{c_i(1-c_i)}(\langle \xi_i \xi_j \rangle - \langle \xi_i \rangle \langle \xi_j \rangle) = \frac{k_{\mathrm{B}} T}{c_i(1-c_i)}\left(\frac{\partial \Omega}{\partial v_i \partial v_j}\right)_{v_i = v \forall i}. \tag{15}$$

A particularly useful quantity is the direct correlation function

$$S_{ij}^{(2)} = (\partial \Omega_e^{\mathrm{CPA}}\{c_i\}/\partial c_i \partial c_j)_{\{\bar{c}_i\}}. \tag{16}$$

For instance, in the disordered state the lattice Fourier transform of α_{ij}, $\alpha(\boldsymbol{k})$, which is measured in diffuse scattering experiments, is given by

$$\alpha(\boldsymbol{k}) = k_{\mathrm{B}} T/(k_{\mathrm{B}} T - c(1-c)\, S^{(2)}(\boldsymbol{k})), \tag{17}$$

where $S^{(2)}$ is the lattice Fourier transform of $S_{ij}^{(2)}$ defined in (16).

3. The self-consistent field Korringa–Kohn and Rostaker coherent potential approximation [SCF-KKR-CPA]

Thus, for very general reasons, detailed in the previous section, we need to solve (13) for the partly averaged charge densities defined in (14). Note that this is a particularly simple example of electrons in disordered potential. It is often referred to as the case of cellular disorder and the CPA is a well-established method for dealing with it (Elliott *et al.* 1974; Ehrenreich & Schwartz 1976). This fact lands a solid foundation to the theory of random substitutional alloys which other random systems such as liquids and glasses do not possess.

Until recently (Kudrnovsky *et al.* 1989), of all the band theory methods like FLAPW, LMTP, KKR only the latter has been adopted for applications to disordered systems. In fact the multiple scattering version of the KKR proved eminently suitable for implementing the basic ideas of the CPA for random substitutional alloys (Gyorffy & Stocks 1979).

In fact the conceptual framework is very simple: sites described by the potential well corresponding to $v^{\alpha}(r-R)$ scatter electrons according to the partial wave scattering amplitudes

$$f_l^{\alpha}(\epsilon) = (1/2\mathrm{i})\,(\exp[\mathrm{i}2\delta_l^{\alpha}(\epsilon)]-1), \tag{18}$$

where $\delta_l^{\alpha}(\epsilon)$ is the usual scattering phase shift which describes the effect of a potential at the origin on the outgoing spherical waves, and we are looking for the electronic structure determined by the fundamental equation of multiple scattering. In a form most suitable for our present purposes it is given by

$$\sum_{L''l} (-\epsilon^{\frac{1}{2}} f_{il}^{-1}\,\delta_{il}\,\delta_{L,L''} - G_{L,L''}(\boldsymbol{R}_i - R_l;\epsilon))\,\tau_{L''L'}^{lj} = \delta_{LL'}\,\delta_{ij}, \tag{19}$$

where L stands for both the polar and azimuthal quantum numbers l and m respectively, $G_{L,L''}(\boldsymbol{R}_i - R_l;\epsilon)$ is the real space KKR structure constant which describes the propagation of free spherical waves of angular momentum L from site to site and $\tau_{LL'}^{ij}(\epsilon)$ is the scattering path operator (Gyorffy & Stocks 1979) which relates an incident wave of angular momentum L to the site j to an outgoing wave of angular momentum L' from the site i. Evidently for the random potential problem at hand $f_{i,l}^{-1} = \xi_i f_{\mathrm{A},l}^{-1} + (1-\xi_i) f_{\mathrm{B},l}^{-1}$.

An important feature of (19) is the separation of the scattering power at the scattering centres described by the scattering amplitudes $f_{i,l}(\epsilon)$, and the geometrical arrangements of such centres which determine the otherwise potential-independent structure constants $G_{LL'}(\boldsymbol{R}_i - \boldsymbol{R}_j;\epsilon)$. Note that the former appears only on the site diagonal part of (19). In the language of tight-binding model hamiltonians, this means the problem at hand corresponds to site diagonal randomness only. Interestingly, this is the case in spite of the fact that the scattering amplitudes $f_{\mathrm{A},l}$ and $f_{\mathrm{B},l}$ can correspond to bands of widely different widths. In other words, remarkably, the KKR-CPA treats on equal footing random alloys of metals with very different band position, band width and hybridization without introducing explicit off-diagonal randomness (Gyorffy & Stocks 1979).

The actual KKR-CPA procedure, based on (19), is very straightforward; we seek an effective (coherent) scattering amplitude which describe the average Greens function. It is the solution of the CPA condition which, for the KKR model, works out to be

$$c_i\,\tau_{LL'}^{\mathrm{A},ij}(\epsilon) + (1-c_i)\,\tau_{LL'}^{\mathrm{B},ij}(\epsilon) = \tau_{LL'}^{\mathrm{C},ij}(\epsilon), \tag{20}$$

where $\tau_{LL'}^{\alpha,ij}(\epsilon)$ is the site diagonal solution of (19) with an α impurity of $\boldsymbol{R}_i$ in the coherent potential lattice and $\tau^{\mathrm{C},ij}$ is the solution when all the sites are described by the effective scattering amplitudes $f_{i,\mathrm{C};L}(\epsilon)$.

Once (20) has been solved for $f_{i,\mathrm{C};L}(\epsilon)$ and $\tau_{LL'}^{\alpha,ij}(\epsilon)$ the partly averaged charge densities are to be calculated using the formula (Gyorffy & Stocks 1979):

$$\bar{n}^{\alpha}(\boldsymbol{r}) = -\frac{1}{\pi}\sum_{L}\int \mathrm{d}\epsilon\, f(\epsilon)\, Z_L^{\alpha}(\boldsymbol{r};\epsilon)\, Z_{L'}^{\alpha}(\boldsymbol{r};\epsilon)\ \mathrm{Im}\ \tau_{LL'}^{\alpha,ij}(\epsilon), \tag{21}$$

where $Z_{\mathrm{L}}^{\alpha}(\boldsymbol{r};\epsilon)$ is the regular, radial solution for an α-type muffin-tin potential well. In principle the above procedure can be iterated to self-consistency. It should be stressed that while the fundamental equation of inhomogeneous KKR-CPA given in (20) is the formal bases for calculating the generalized grand potential $\Omega_e^{\mathrm{CPA}}(\{c_i\})$ it can be solved only in the homogeneous limit where $c_i = \bar{c}\forall i$. Nevertheless, (20) can be used to derive computationally tractable expressions (Gyorffy & Stocks 1983) for the direct correlation function given in (16).

Before drawing this brief summary of our theory to a close, we comment on the nature of the electronic structure as described by the KKR-CPA.

In general, for a disordered system the wave vector $\boldsymbol{k}$ is not a good quantum number. Moreover, in the generic case, like a liquid or a glass, on the average the system is translationally invariant and hence the relevant wave vectors comprise all of $\boldsymbol{k}$-space. Namely the Brillouin zone is of infinite extent. Under these circumstances one is tempted to abandon the use of $\boldsymbol{k}$-space altogether in favour of real space methods like cluster or supercell calculations (Zunger *et al.* 1990). However, the cellular disorder of crystalline random alloys is an intermediate case between ordered crystals and the topologically disordered systems mentioned above. Here, the ensemble of configurations is invariant under the discrete set of translations which defines the underlying lattice and this introduces a periodicity in $\boldsymbol{k}$-space. That is to say, there are Brillouin zones of finite extent and the phase space to be considered is reduced to one of these. Of course, $\boldsymbol{k}$ is still not a good quantum number since the translational symmetry applies only on the average. Nevertheless, it turns out to be a surprisingly useful parameter. To make the best use of it one defines the Bloch spectral function

$$A_{\mathrm{B}}(\boldsymbol{k};\epsilon) = -\frac{1}{\pi}\sum_j \mathrm{e}^{\mathrm{i}\boldsymbol{k}(\boldsymbol{R}_j-\boldsymbol{R}_i)} \int_{\Omega_i} \mathrm{d}r^3 \,\mathrm{Im}\, \langle G(r+R_j, r+R_i;\epsilon)\rangle, \tag{22}$$

where the integral is over the ith unit cell whose volume is Ω_i. Within the KKR-CPA it is quite straightforward to evaluate the Bloch spectral function, which is automatically periodic in $\boldsymbol{k}$ space, and it gives the most complete account of the electronic states of a random alloy.

For an ordered system $A_{\mathrm{B}}(\boldsymbol{k};\epsilon)$ is given by a set of delta function peaks at $\epsilon = \epsilon_{\boldsymbol{k}}$, v, the Bloch energy eigenvalues for the band index v. Naturally, for disordered systems these peaks broaden out and their width can be interpreted as the inverse lifetime of the state specified by $\boldsymbol{k}$ and v. As an example we show, in figure 1, the Bloch spectral function at the Fermi energy ϵ_{F} in the ΓXWKWXΓ plane of the Brillouin zone for the very interesting $\mathrm{Cu_{76}Pd_{24}}$ alloy system. Clearly, there is a Fermi surface well defined on the scale of its linear dimensions. Furthermore, it has interesting features, like the pronounced flat sheet perpendicular to the ΓK direction, and these can be measured in 2^{d} angular correlation of (position) annihilation radiation (ACAR) experiments (Berko 1979).

4. The configurationally averaged total energy

The above is one of the first useful results that comes out of a SCF-KKR-CPA calculation. The formula for it is a non-trivial, combined consequence of DET in the LDA and the stationarity property of the CPA. Nevertheless, it takes the following simple form (Johnson *et al.* 1990):

$$\bar{E} = cE_J[\bar{n}^{\mathrm{A}}, \bar{n}_0] + (1-c)\,E_J[\bar{n}^{\mathrm{B}}, \bar{n}_0], \tag{23}$$

where E_J is the total energy functional derived by Janak (1974) for ordered systems with the crystal potential in the muffin-tin form, $\bar{n}^{\mathrm{A}}(\boldsymbol{r})$ and $\bar{n}^{\mathrm{B}}(\boldsymbol{r})$ are the partly averaged local charge densities defined in (14), $\bar{n}(\boldsymbol{r})$ is the fully averaged charge

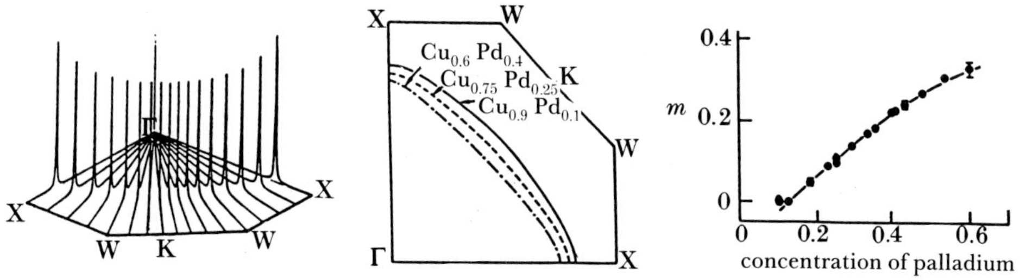

Figure 1. (*a*) The Bloch spectral function $A(\boldsymbol{k};\epsilon_F)$ at the Fermi energy ϵ_F in the ΓXWKWXΓ plane of the Brillouin zone for the $Cu_{0.75}Pd_{0.25}$ alloy. (*b*) Evolution of the calculated Fermi surface with concentration *c*. (*c*) Variation of the incommensurability $m = 2(\sqrt{2}-2k_F(011))$ with concentration.

Figure 2

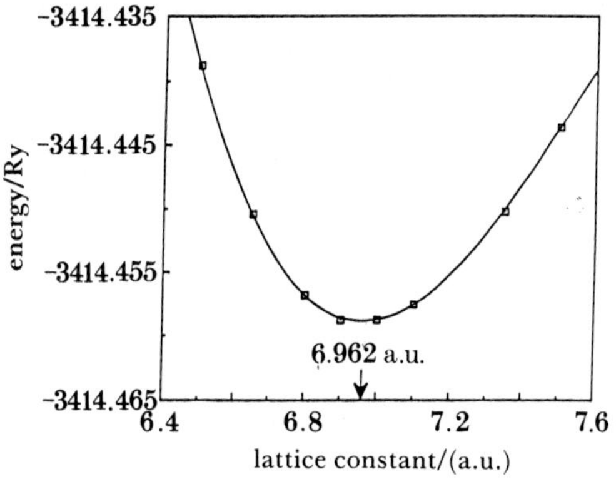

Figure 3

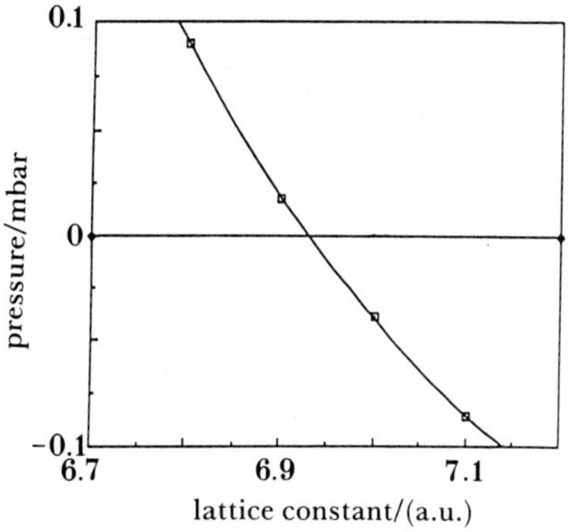

Figure 2. The variation of the configurationally averaged total energy (in rydbergs) with lattice constants (in atomic units) for the FCC $Cu_{0.5}Zn_{0.5}$ alloy. A cubic least-squares fit gives a minimum energy of -3414.45 Ry at 6.93 a.u. and a bulk moduli of 1.41 mbar (1.41×10^{11} Pa).

Figure 3. The variation of the alloy pressure (in rydbergs) with lattice constant for FCC $Cu_{0.5}Zn_{0.5}$. The zero pressure point is at $a_P = 6.93$ a.u. and the bulk moduli is 1.41 mbar (1.41×10^{11} Pa).

density and $\bar{n}_0$ is the same but in the interstitial region. A similar formula applies for the pressure (Johnson *et al.* 1990). A remarkable feature of (23) is that, due to the use of the CPA, *E* has similar variational characteristics as the total energy in the ordered state. Namely,

$$\delta\bar{E}/\delta\bar{n}^{A}(\boldsymbol{r}) = 0; \quad \delta\bar{E}/\delta\bar{n}^{B}(\boldsymbol{r}) = 0. \tag{24}$$

This property is, certainly, one of the principal reasons for the success of the SCF-KKR-CPA in predicting the total energy as a function of concentration.

In figure 2 we present our calculated $\bar{E}$ as a function of the lattice constant for the FCC $Cu_{0.50}Zn_{0.50}$ random alloy. The minimum at $a_E = 6.96$ a.u.† is within 5% of the experimentally determined equilibrium value. Moreover, the calculated pressure against '*a*' curve, shown in figure 3, crosses zero at $a_P = 6.93$ a.u., which is in very satisfactory agreement with a_E determined from the total energy calculation. Thus the consistency and accuracy of these calculations are comparable with those achieved in similar calculations in ordered systems (Maruzzi *et al.* 1978).

Repeating the above calculations for various concentrations yields the lattice parameter against concentration curve shown in figure 4. A particularly pleasing

† 1 a.u. $\approx 5.3\times10^{-11}$ m.

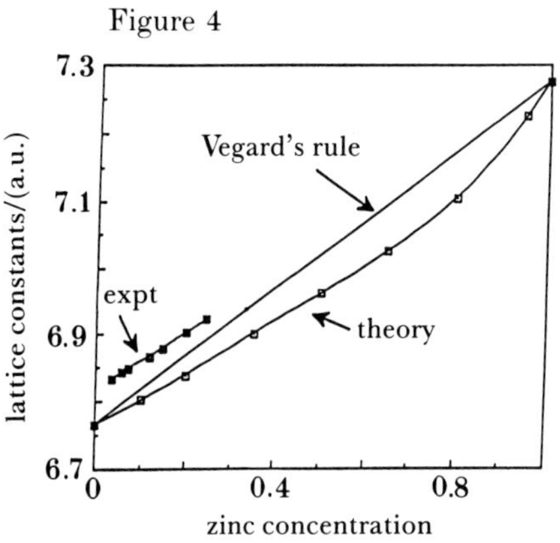

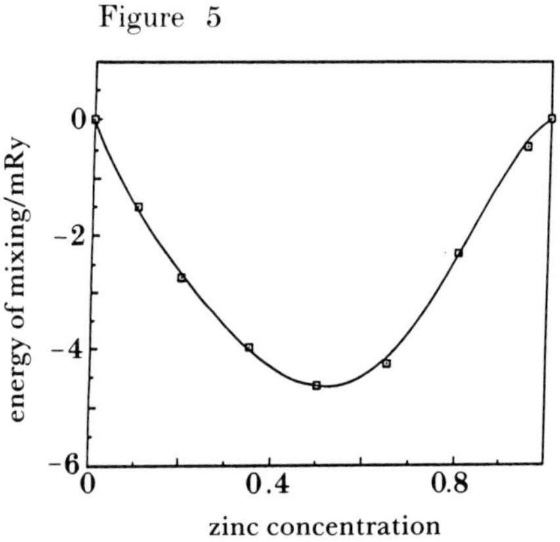

Figure 4. The variations of the lattice constant a with concentration as calculated by the SCF-KKR-CPA. For comparison we also show the prediction of Vegard's rule.

Figure 5. The calculated heats of mixing at various concentrations for the FFC Cu_cZn_{1-c} alloys.

feature of the generally good agreement between the theory and experiment is the fact that $(\partial a/\partial c)_{c=0} = 0.0036$ per at. % Zn which deviates from the prediction of the Vegard's rule of 0.0052 per at. % Zn but agrees with the experimental result of 0.0036 per at. % Zn. While it should be stressed that the experimental determination was at $T = 300$ K the thermal expansion coefficient for this alloy is small and hence the agreement is real at the 0.5% level.

Another quantity of interest is the heat of mixing given by

$$\Delta E^{\text{mix}} = \bar{E} - cE^{\text{A}} - (1-c)\,E^{\text{B}}, \tag{25}$$

where E^{A} and E^{B} are the ground state energies of the pure A and pure B metals, respectively, on the alloy lattice. Our calculated values of ΔE^{mix} at various concentrations are shown in figure 5. The negative sign of ΔE^{mix} implies a tendency to order which is consistent with the rather complicated phase diagram which features various order phases (Hansen 1958). Indeed the fact that ΔE^{mix} is not a parabolic function of c suggests such a phase diagram.

More details concerning these calculations are found in the recent paper of Johnson *et al.* (1990).

5. Order–disorder transformations

One of the most intriguing complex of ordering phenomena takes place in the Cu rich $Cu_{1-c}Pd_c$ alloys (Oshima & Watanabe 1976; Ceder *et al.* 1989). For instance, in the high-temperature disordered phase diffuse scattering experiments find in $\alpha(\boldsymbol{q})$ [110] superlattice peaks, split into four peaks. This implies a tendency to form concentration waves incommensurate with the underlying lattice (Sato & Toth 1965). Moreover, these peaks move apart as c changes from 0.12 to 0.4, where they disappear (Oshima & Watanabe 1976). The first success of the first-principle mean field theory described in §2 was to give a microscopic explanation of these observations (Gyorffy & Stocks 1983; Gyorffy *et al.* 1989). In fact, the calculations identified the parallel flat sheets of the Fermi surface, whose presence can be deduced from the Bloch spectral function displayed in figure 1, as the electronic driving mechanism behind the ordering process. Examples of the calculated four peaked $\alpha(\boldsymbol{k})$ are shown in figure 6.

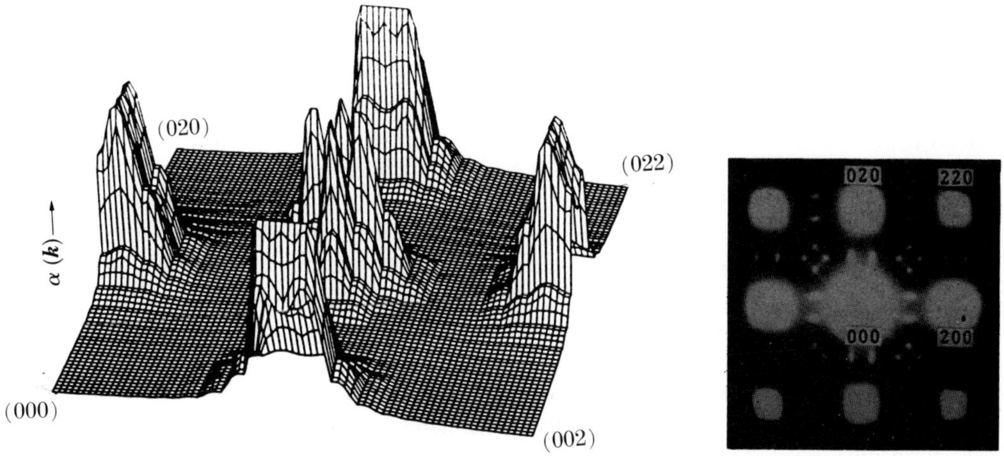

Figure 6. The concentration–concentration correlation function $\alpha(\boldsymbol{k})$ (the Warren–Cowley short-range order parameter) as calculated in the first-principles mean field theory based on the SCF-KKR-CPA in the plane containing the reciprocal lattice points 000, 020, 022, 002 for various $Cu_{1-c}Pd_c$ alloys.

Figure 7

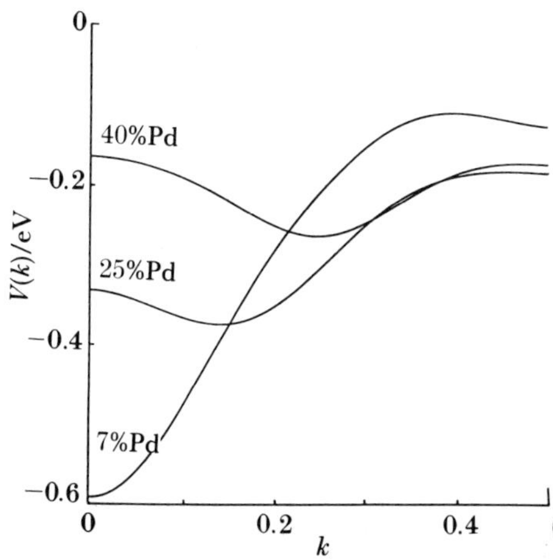

Figure 8

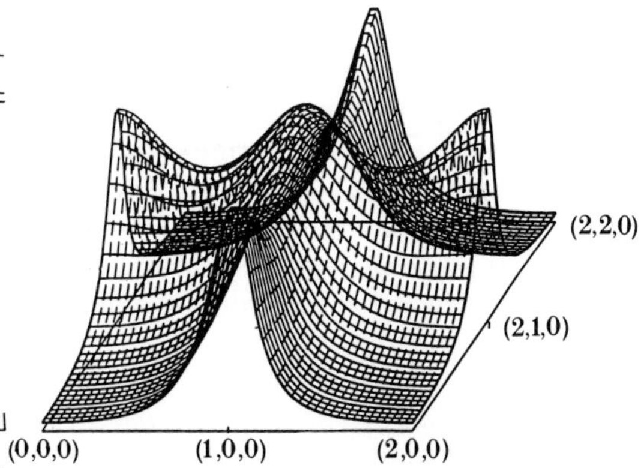

Figure 7. The calculated effective pair interaction parameter $\boldsymbol{v}(\boldsymbol{k})$ along the XW segment (k, 10) in the first Brillouin zone for 7%, 25% and 40% Pd in Cu_cPd_{1-c} alloys.

Figure 8. The calculated Warren–Cowley short-range order parameter $\alpha(\boldsymbol{k})$ in the $k_z = 0$ plane for $T = 1.1\ T_0$, where T_0 is the theoretical ordering temperature. The peaks at the X-points indicate that the alloy will order into a $Q = (1,0,0)$ concentration wave ($L1_0$) state for $T < T_0$.

As is well known, with certain caviats notwithstanding, the direct correlation function $S^{(2)}(\boldsymbol{k})$, from which $\alpha(\boldsymbol{k})$ is calculated according to (7), can be regarded as the lattice Fourier transform of an effective pairwise, interchange, potential $\tilde{v}(\boldsymbol{k})$ [$\equiv v^{AA}(\boldsymbol{k}) + v^{BB}(k) - 2v^{AB}(\boldsymbol{k})$]. To display the subtle ways the ordering tendency changes in these alloys we plot $S^{(2)}(\boldsymbol{k};c) = \tilde{v}(\boldsymbol{k};c)$ for various concentrations in figure 7. Recently, using these curves, in figure 7 and the mean field theory Cedar *et al.* (1991) have constructed the full, very complex, low-temperature phase diagram featuring lock in transitions into high-order commensurate phases, in good qualitative agreement with experiments.

Another recent application of the method is to the Pt_c–Ni_{1-c} alloy system. These alloys represent a serious problem for theories based on simple tight-binding models of the electronic structure because they predict a trend according to which alloys with nearly half-filled d-band order and those for which ϵ_F falls near the bottom or the top of the effective d-band phase separate (Heine & Samson 1983; Teleglia & Ducastelle 1987). Experimentally Pt_c–Ni_{1-c} is found to order, contradicting the above trend governed by the band filling or electron per atom ratio e/a.

Surprisingly, the first principles mean field theory based on the SCF-KKR-CPA predicts ordering in agreement with experiments. We show $S^{(2)}(k)$ in the $k_z = 0$ plane in figure 8. The sharp rise of $S^{(2)}(\boldsymbol{k})$, from the Γ point to the zone boundary, can be interpreted as a robust tendency to order in agreement with the experiments of Dahmani *et al.* (1985). The reason for this dramatic difference between the tight-binding model calculations and the SCF-KKR-CPA-based theory is very interesting. Analysis of our results suggest that ordering in this system is due to the size effect which overpowers the competing e/a effect. In other words our first principles calculation correctly reproduces the metallurgical rule of thumb according to which alloys of large and small atoms order (on an FCC lattice into an $L1_2$ structure).

In the language of tight-binding model hamiltonians big and small atoms, on the same lattice, give rise to large and small overlap integrals and hence bandwidths. This in turn causes off-diagonal randomness which, in general, is difficult to treat. The key to the success of the first principles calculation is the SCF-KKR-CPA method which treats site-diagonal, site-off-diagonal and hybridization randomness on equal footing.

In summary we stress that our first principles mean field theory, while subject to limitations due to the neglect of certain correlations, treat all the classic Hume-Rothery factors which determine the state of compositional order, namely electron per atom ratio, size effect and electronegativity, on equal footing and without adjustable parameters. The most reassuring aspect of the above Pt_c–Ni_{1-c} example is that it appears to be an adequate theory of the relative significance of these factors even when they drive the ordering and clustering tendency in the opposite direction.

References

Berko, S. 1979 In *Electrons in disordered metals and metallic surfaces* (ed. P. Phariseau, B. L. Gyorffy & L. Scheire) (Plenum Press ASI Series, Physics B **42**); 1989 *Momentum distributions* (ed. R. N. Silver & P. E. Sokol). Plenum.

Cedar, G., de Fontaine, D., Dreysse, H., Nicholson, D. M., Stocks, G. M. & Gyorffy, B. L. 1991. (In preparation.)

Ceder, G., de Graef, H., Delacy, L., Kulik, J. & de Fontain, D. 1989 *Phys. Rev.* B **39**, 381.

de Fontain, D. 1979 *Solid state physics* (ed. H. Ehrenreich, F. Seitz & D. Turnbull), vol. 34. Academic Press.

Dahmani, C. E., Cadeville, M. C. & Moran-Lopez, X. 1985 *Phys. Rev. Lett.* **55**, 1208.

Ehrenreich, H. & Schwartz, L. 1976 *Solid state physics* (ed. H. Ehrenreich, F. Seitz & D. Turnbull). Academic Press.

Elliott, R. J., Krumhansl, J. A. & Leath, P. L. 1974 *Rev. mod. Phys.* **46**, 465.

Gyorffy, B. L. & Stocks, G. M. 1979 *Electrons in disordered metals and at metallic surfaces* (ed. P. Phariseau, B. L. Gyorffy & L. Scheire). Plenum Press ASI Series, Physics B **42**.

Gyorffy, B. L. & Stocks, G. M. 1983 *Phys. Rev. Lett.* **50**, 374.

Gyorffy, B. L., Johnson, D. D., Pinski, F. J., Nicholson, D. M. & Stocks, G. M. 1989 *Alloy phase stability*. ASI Series E, vol. 163. Kluwer.

Hansen, M. 1958 *Constitution of binary alloys*. New York: McGraw-Hill.
Heine, V. & Samson, J. 1983 *J. Phys.* F **13**, 2155.
Hume-Rothery, W. & Coles, B. R. 1969 *Atomic theory for students of metallurgy*. The Institute of Metals.
Janak, J. F. 1974 *Phys. Rev.* **39**, 3985.
Johnson, D. D., Nicholson, D. M., Pinski, F. J., Gyorffy, B. L. & Stocks, G. M. 1990 *Phys. Rev.* B **41**, 9701.
Kchaheturyan, A. G. 1986 *Theory of structural transformations in solids*. Wiley.
Kohn, W. & Vashishta, P. 1983 *Theory of the inhomogeneous electron gas* (ed. S. Lundquist & N. H. March). Plenum Press.
Kudrnovsky, J., Drchal, V., Sob, M., Christensen, N. E. & Anderon, O. K. 1989 *Phys. Rev.* B **40**, 10029.
Maruzzi, V. L., Janak, J. F. & Williams, A. R. 1978 *Calculated electronic properties of metals*. New York: Pergamon Press.
Mott, N. F. & Jones, H. 1936 *The theory of the properties of metals and alloys*. Clarendon Press.
Oshima, K. & Watanabe, D. 1976 *Acta crystallogr.* A **32**, 883.
Sato, H. & Toth, R. S. 1965 *Alloying behaviour and effects in concentrated solid solutions* (ed. B. T. Massalski). Gordon and Breach.
Schwartz, L. & Sigga, E. 1972 *Phys. Rev.* B **5**, 383.
Stocks, G. M. & Winter, H. 1984 In *Electronic structure of complex systems* (ed. P. Phariseau & W. M. Tennermann). ASI Series B113. Plenum Press.
Teleglia, G. & Ducastelle, F. 1987 *J. Phys.* F **50**, 374.
Ziman, J. M. 1979 *Models of disorder*. Cambridge University Press.
Zunger, A., Wei, S.-H., Ferreira, L. G. & Bernard, J. E. 1990 *Phys. Rev. Lett.* **65**, 353.

Discussion

H. Rafii-Tabar (*Department of Materials, Oxford University, U.K.*). Is there a simple relationship between the A–B interaction on the one hand and the A–A and B–B interactions on the other hand in Professor Gyorffy's calculations?

B. L. Gyorffy. This is a first principles electronic theory in which there are no pair potentials as such. The direct correlation function $S^{(2)}(\boldsymbol{k})$ may, however, be interpreted (in the mean spherical approximation) as the interchange energy $v(\boldsymbol{k}) = v^{AA}(\boldsymbol{k})+v^{BB}(\boldsymbol{k})-2v^{AB}(\boldsymbol{k})$, which occurs in pair potential models. But even then one can not identify anything in our theory which could be called as $v^{AA}(\boldsymbol{k})$ or $v^{BB}(\boldsymbol{k})$ or $v^{AB}(\boldsymbol{k})$ on their own.

J. B. Pendry (*Imperial College, London, U.K.*). Although CPA is a mean field theory can it be applied to photoemission and transport properties which involve higher-order correlation functions?

B. L. Gyorffy. Yes. The CPA provides a prescription for higher-order averages such as $\langle GG \rangle$ and $\langle GGG \rangle$ without replacing them by products of $\langle G \rangle$. In other words there are CPA vertex corrections. In fact first principles (KKR-CPA) calculations have been done for transport problems photoemission problem and positron annihilation problem, all of which involve such higher-order averages.

N. W. Ashcroft (*Cornell University, U.S.A.*). Evidently, size differences in atoms are being accommodated only insofar as they are manifested in the electronic (band) part of the problem. Yet as one knows from the liquid equivalent of this problem, size differences give rise to a significant entropic component to the overall free energy (i.e. going beyond the (ln C type of term that is being used). This size difference is also important in the known failures of Vegard's Law (most binary systems deviate from this rule). Given this, is it possible to improve on the entropic contributions, as they enter the theory?

B. L. GYORFFY. What I presented is an all electron theory and hence, as a matter of principle, no explicit consideration of atomic size enters the problem. The effect whose liquid state analogue Professor Ashcroft referred to would come into the present discussion if and when the size of the unit cell is allowed to fluctuate as the occupance changes I called these strain fluctuations (they are driven by the Kanzaki forces) and said that for the time being they are not taken into account. Clearly, their contribution to the free energy would have an entropic contribution. However, in the case of the Pt_cNi_{1-c} they would surely favour ordering and therefore would reinforce the ordering tendency we found.

V. KUMAR (*Materials Science Division, Indira Gandhi Centre for Atomic Research, Kalpakkam, India*). In the tight-binding model of disordered alloys, effects of local environments could be included reasonably well by treating a configurational average over a cluster of atoms and the single site coherent potential. Has such an attempt been made for KKR-CPA and if so how well does it work?

B. L. GYORFFY. Formally the KKR-CPA is the same as its tight-binding version and new developments can be readily recast into the language of KKR-CPA. The difficulty is the computational implementation. There are a few cluster KKR-CPA studies in the literature (Gonis *et al.* 1983) but not many.

L. M. FALICOV (*University of California, Berkeley, U.S.A.*). The CPA is now 24 years old. It was received originally as the alloy theory that could solve all problems in the field. However, no step beyond the single-site (concentration) averaging has been successfully completed yet. Was the optimism misplaced? Is there a chance that the impasse will be overcome?

B. L. GYORFFY. That CPA 24 years ago was for a simple tight-binding model with the uncontrolled parameters and no reference to the thermodynamic state of the nuclei. The theory I presented is a full theory, albeit mean field and single site, of the nuclei and electron system without adjustable parameters. Note that we have successfully calculated the pair correlation functions, $\langle \xi_i \xi_j \rangle - \langle \xi_i \rangle \langle \xi_j \rangle$ for a variety of real systems. Such considerations were not even dreamed of in connection with the CPA in those days. As to the technical questions of whether there is a natural, universally expected approximation which goes beyond the single site CPA the short answer is that there is not. Nevertheless, there are useful recipes like cluster CPA's and direct space methods based on the inverse Monte-Carlo algorithms of significant structure but most of these are as universally accepted as the good old CPA.

Additional reference

Gonis, A., Butler, W. H. & Stocks, G. M. 1983 *Phys. Rev. Lett.* **50**, 1482.

Surface properties in an external electric field

By J. E. Inglesfield

Institute for Theoretical Physics, Catholic University of Nijmegen, Toernooiveld, NL-6525 ED Nijmegen, The Netherlands

This paper reviews recent calculations on the effect of an external electric field on surface electronic properties, in particular using the embedding method for solving the Schrödinger equation at the surface. The shape of the screening charge and its field dependence are discussed, and the results are compared with experiments in which the image plane is determined. The force on the surface atoms in the field is given in terms of an effective charge, which also determines the work-function variation with surface displacements. This relationship can lead to a surface instability if the effective charge is big enough.

1. Introduction

In this paper I discuss recent calculations on the effect of an external field on a metal surface. Classically we know what happens: the field is perfectly screened, with a surface charge density of $\mathscr{E}/4\pi$. However, the microscopic details of the screening are important for many problems in surface science (Inglesfield 1990): in the field emission and field ion microscopes, and the scanning tunnelling microscope, an electric field is applied to the surface, and in electrochemistry there is a strong field across the Helmholtz double layer at the surface of the electrode (Kolb *et al.* 1981). The centre of gravity of the screening charge (averaged across a plane parallel to the surface) corresponds to the reference plane for the image potential (Lang & Kohn 1973), of current interest given the data available from inverse photoemission experiments on the energies of the image-potential-induced surface states (Smith *et al.* 1989). Nonlinear effects in screening, manifesting themselves as a change in the shape of the screening charge with field strength, show up in second harmonic generation when laser light is reflected from the surface (though our results for the screening of a static field can only be applied in the low-frequency limit (Weber & Liebsch 1987*a*)).

The atoms themselves respond to an external field; the net force on the metal is, after all, $\mathscr{E}^2/8\pi$ per unit area. But once again a detailed description of the forces on the atoms must be important for understanding the process of field evaporation where surface atoms are stripped off by a strong field: on the W (001) surface, for example, preferential field evaporation occurs in field ion experiments, removing alternate atoms to give a $(\sqrt{2}\times\sqrt{2})R45°$ vacancy structure (Melmeed *et al.* 1979). Computations show that an external field can induce a (1×2) reconstruction of the Ag (110) surface (Fu & Ho 1989), suggesting that the effect of alkali adsorbates in driving this reconstruction is due to an excess surface charge resulting from alkali ionization, analogous to the screening charge in the external field. Another example of surface atoms responding to an external field is in electron energy loss (EELS) experiments, where surface phonons are excited by low-energy electrons: in the dipole scattering régime this can be discussed in terms of effective charges on the surface atoms.

In §§2 and 3 I review calculations of the screening charge for different surfaces, and use this to discuss experimental results on image potentials and second harmonic generation. The calculations involve the self-consistent solution of the electronic Schrödinger equation with surface geometry, and I describe in particular the embedding method (Inglesfield & Benesh 1988) which can be used to find the electronic states at the surface of a semi-infinite solid as an alternative to the more widely used slab geometry (Krakauer *et al.* 1979). To discuss forces on surface atoms in the presence of a field I discuss the concept of effective charges in §4, and then use this in §5 to discuss a surface instability driven by effective charges and its possible application to O chemisorbed on Cu (001).

2. Surface electronic structure in an external field

The first calculations of the screening of an external field at a surface were carried out by Lang & Kohn (1970), for free-electron-like metal surfaces modelled by semi-infinite jellium. Like (almost) all subsequent surface calculations, this was carried out within the mean-field framework of density functional theory; in addition to the Hartree and external potentials, an electron feels an exchange-correlation potential to describe the effects of exchange and the correlated motion of the electrons. The exchange-correlation potential V_{xc} is local and energy independent, and it can be found quite accurately using the local density approximation – $V_{xc}(\boldsymbol{r})$ is taken as the exchange-correlation potential of a uniform electron gas with the density of electrons at $\boldsymbol{r}$. As the Hartree and exchange-correlation potentials depend on the electron density, this has to be found self-consistently; in the presence of an external field, the screening charge drops out automatically in the process of iterating to self-consistency. Solving the one-electron Schrödinger equation itself is relatively straightforward for the jellium surface, because the potential is one-dimensional and it is easy to match the wavefunctions in the surface region onto the bulk solutions (Lang & Kohn 1970). Jellium calculations are still extremely useful in studying surface response, especially the frequency-dependent response for which they are the only feasible calculation at present, and Liebsch and his co-workers have discussed second harmonic generation in this way (Weber & Liebsch 1987*a*, *b*; Liebsch & Schaich 1989). However, we shall see in §3 that there are very significant differences in (static) screening when the atoms are taken into account.

In general it is more difficult to solve the Schrödinger equation at the surface than in the bulk, because of the absence of translational symmetry in the direction perpendicular to the surface: only the Bloch wavevector $\boldsymbol{K}$ parallel to the surface is a good quantum number. In principle it can be done using the generalization of Lang & Kohn's (1970) wavefunction matching to the three-dimensional case. As Heine has shown (1963), there is a one-to-one relationship between the solutions of the bulk Schrödinger equation, travelling towards or away from the surface at energy E and with wavevector component $\boldsymbol{K}$, and the surface reciprocal lattice vectors $\boldsymbol{G}$. This means that there are exactly the right number of bulk solutions for matching onto a surface solution in amplitude and derivative, assuming that these are expanded as a Fourier series in $\boldsymbol{G}$ over some interface plane separating the bulk and surface regions. This result is very important conceptually for understanding surface electronic structure, but explicit wavefunction matching is rarely used in practise (apart from the pioneering work of Appelbaum & Hamann (1972, 1973) on Na (001) and Si (111)). The usual approach in surface calculations is to use slab geometry, so that one is dealing with a system of finite thickness (typically five atomic layers thick) for which ordinary basis set methods can be used. Because quantities involving sums over states like charge density and energy are very local, the presence of the second surface of the slab does not matter for cohesive properties. However, for a precise description of individual electronic states, slab geometry is not adequate: in slab geometry, all the states at fixed $\boldsymbol{K}$ are discrete, and there is no real distinction between the discrete surface states localized at the surface, and bulk states bouncing off the surface. It is also unsatisfactory, perhaps, to solve the Schrödinger equation for two surfaces separated by a piece of bulk material, when all we want are the surface properties.

The 'surface embedded Green function method' (SEGF) provides a practical scheme for solving the Schrödinger equation for the surface of a real semi-infinite solid. As in wavefunction matching, this method assumes that the semi-infinite solid can be divided into two regions: region I which is the real surface region, say the top layer or two of atoms and the vacuum, and region II where an electron feels essentially the bulk potential. We then solve the Schrödinger equation explicitly only in the surface region, adding on to the hamiltonian a complex, energy-dependent embedding potential to describe the scattering of the electrons by the substrate region II.

To obtain the embedding potential (Inglesfield 1981), we start from the variational principle, with an arbitrary trial function $\phi(\boldsymbol{r})$ defined in region I. This trial function can in principle be extended into region II by finding the solution of the Schrödinger equation in this region at some trial energy ϵ, which matches onto ϕ in amplitude over the interface S between the two regions; let us call this ψ. The expectation value of the hamiltonian in the whole system is then:

$$E = \left[\int_{\mathrm{I}} \mathrm{d}^3\boldsymbol{r}\,\phi^*(\boldsymbol{r})\,H\phi(\boldsymbol{r}) + \epsilon\int_{\mathrm{II}} \mathrm{d}^3\boldsymbol{r}\,|\psi|^2 + \frac{1}{2}\int_{\mathrm{S}} \mathrm{d}^2\boldsymbol{r}_s\,\phi^*\frac{\partial\phi}{\partial n_s} - \frac{1}{2}\int_{\mathrm{S}} \mathrm{d}^2\boldsymbol{r}_s\,\phi^*\frac{\partial\psi}{\partial n_s}\right] \times\left[\int_{\mathrm{I}} \mathrm{d}^3\boldsymbol{r}\,|\phi|^2 + \int_{\mathrm{II}} \mathrm{d}^3\boldsymbol{r}\,|\psi|^2\right]^{-1}. \quad (1)$$

The surface integrals in (1) come from the discontinuity in the normal derivative of the trial function across S. The fundamental principle of embedding is that the integrals through the substrate region can be eliminated, using the Green function for the bulk crystal satisfying the boundary condition on S that:

$$\partial G_0(\boldsymbol{r}_s, \boldsymbol{r}')/\partial n_s = 0. \quad (2)$$

The inverse of G_0 over S is the generalized logarithmic derivative, relating the derivative of a solution of the Schrödinger equation in II to the amplitude:

$$\frac{\partial\psi(\boldsymbol{r}_s)}{\partial n_s} = -2\int_{\mathrm{S}} \mathrm{d}^2\boldsymbol{r}'_s\,G_0^{-1}(\boldsymbol{r}_s, \boldsymbol{r}'_s)\,\psi(\boldsymbol{r}'_s). \quad (3)$$

Substituting (3) into (1) and making use of a relationship between normalization integrals in II and the energy derivative of $\partial\psi/\partial n_s$ we finally obtain:

$$E = \left[\int_{\mathrm{I}} \mathrm{d}^3\boldsymbol{r}\,\phi^*(\boldsymbol{r})\,H\phi(\boldsymbol{r}) + \frac{1}{2}\int_{\mathrm{S}} \mathrm{d}^2\boldsymbol{r}_s\,\phi^*\frac{\partial\phi}{\partial n_s} + \int_{\mathrm{S}} \mathrm{d}^2\boldsymbol{r}_s\int_{\mathrm{S}} \mathrm{d}^2\boldsymbol{r}'_s\,\phi^*(\boldsymbol{r}_s)\left(G_0^{-1} - \epsilon\frac{\partial G_0^{-1}}{\partial\epsilon}\right)\phi(\boldsymbol{r}'_s)\right] \times\left[\int_{\mathrm{I}} \mathrm{d}^3\boldsymbol{r}\,|\phi|^2 + \int_{\mathrm{S}} \mathrm{d}^2\boldsymbol{r}_s\int_{\mathrm{S}} \mathrm{d}^2\boldsymbol{r}'_s\,\phi^*(\boldsymbol{r}_s)\frac{\partial G_0^{-1}}{\partial\epsilon}\phi(\boldsymbol{r}'_s)\right]^{-1}. \quad (4)$$

This expression gives us the expectation value of E purely in terms of the trial function ϕ defined in region I, and values of the embedding potential G_0^{-1} over S.

To find the actual wavefunction in the surface region we can now expand ϕ in terms of any suitable set of basis functions (we actually use linearized augmented plane waves, LAPWS (Krakauer *et al.* 1979)):

$$\phi(\boldsymbol{r}) = \sum_i a_i \chi_i(\boldsymbol{r}), \tag{5}$$

and minimizing E in (4) gives the matrix equation for the coefficients:

$$\sum_j \left[H_{ij} + (G_0^{-1})_{ij} + (E-\epsilon)\frac{\partial (G_0^{-1})_{ij}}{\partial \epsilon} \right] a_j = E \sum_j S_{ij} a_j. \tag{6}$$

The matrix elements are given by:

$$\left.\begin{aligned} H_{ij} &= \int_{\mathrm{I}} \mathrm{d}^3\boldsymbol{r}\chi_i^* H\chi_j + \frac{1}{2}\int_{\mathrm{S}} \mathrm{d}^2\boldsymbol{r}_s \chi_i^* \frac{\partial \chi_j}{\partial n_s}, \\ (G_0^{-1})_{ij} &= \int_{\mathrm{S}} \mathrm{d}^2\boldsymbol{r}_s \int_{\mathrm{S}} \mathrm{d}^2\boldsymbol{r}'_s \chi_i^* G_0^{-1} \chi_j, \\ S_{ij} &= \int_I \mathrm{d}^3\boldsymbol{r}\chi_i^* \chi_j. \end{aligned}\right\} \tag{7}$$

H_{ij} is the matrix element of the hamiltonian in region I, with the additional surface integral which ensures hermiticity. $(G_0^{-1})_{ij}$ is the matrix element of the embedding potential, evaluated at energy ϵ, and the energy derivative term in (6) is the first-order correction to give it at the right energy. The embedding potential is, in fact, a *pseudopotential* (Heine 1970) replacing the whole of region II: the relationship between the energy derivative of the embedding potential and the normalization of the wavefunction in the region of space which it is replacing is, of course, familiar from standard pseudopotential theory (Shaw & Harrison 1967).

In surface applications of the embedding method (Inglesfield & Benesh 1988) it is more convenient to evaluate the surface Green function rather than individual wavefunctions, because at a particular wavevector $\boldsymbol{K}$ the bulk states hitting the surface form a continuum. From the Green function we can immediately find the local density of states, which when integrated over energies up to E_{F} gives the charge density. To go to self-consistency we must solve Poisson's equation in the surface region with this charge density, with the boundary condition that the potential over interface S equals the bulk potential. Deep in the vacuum the boundary condition on the Hartree potential is that $\mathrm{d}V/\mathrm{d}z$ equals the applied electric field $\mathscr{E}$, and this is the only point at which the field enters the calculation (Aers & Inglesfield 1989).

3. Screening charge at Ag (001) and Al (001) surfaces

When an electric field is applied to a surface, the most striking feature of the resulting screening charge distribution is that it is located on top of the surface atoms, so that the field barely penetrates the solid. Figure 1 shows the screening charge at Ag (001) (Aers & Inglesfield 1989), calculated using the SEGF embedding method,

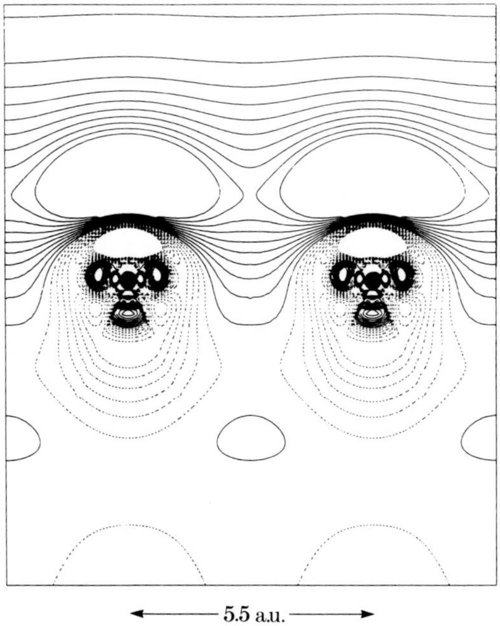

Figure 1. Screening charge at Ag (001) with field $\mathscr{E} = 0.01$ a.u. The plane passes through atoms in the top layer (indicated by heavy dots), and between atoms in the second layer; vacuum is at the top of the figure, and the bulk solid at the bottom. Solid lines are contours of decreased electron density, and dashed lines increased density.

with an applied field of $\mathscr{E} = +0.01$ a.u.† (5×10^9 V m^{-1}) (my convention is that a positive field repels electrons from the surface). There is atomic structure apparent in the screening charge, with polarization effects inside the ion cores; however, the main effect of the ion cores is to exclude the screening charge, which bends over the tops of the cores into the region between the atoms. The corresponding change in potential due to the application of the field is shown in figure 2; we see the equispaced potential contours outside the solid, but we also see how effective the screening charge is in excluding the field. This is relevant to field evaporation, as the force on an atom in an applied field is just the screened field at the nucleus, from the Hellmann–Feynman theorem; we explore this further in §4.

The shape of the screening charge is field-dependent, corresponding to nonlinear screening. In figure 3 we show the planar averaged screening charge as a function of z, the distance from the geometrical surface (where the solid is chopped in two), for fields of ± 0.02 a.u. at Ag (001). We see that an increasing positive field tends to push the screening charge into the solid, whereas an increasing negative field tends to pull it out. From our results we find that the centre of gravity of this screening charge distribution, as a function of $\mathscr{E}$, can be quite well fitted by the straight line:

$$z_0 = -0.97 + 8.83\mathscr{E} \quad \text{(in atomic units).} \tag{8}$$

The zero-field value of z_0 is the electrostatic origin of the surface, from which the asymptotic form of the image potential should be measured; so at Ag (001) the image plane lies at -0.97 a.u., on the *vacuum* side of the geometrical surface.

† 1 a.u. $\approx 5.3 \times 10^{-11}$ m.

Figure 2

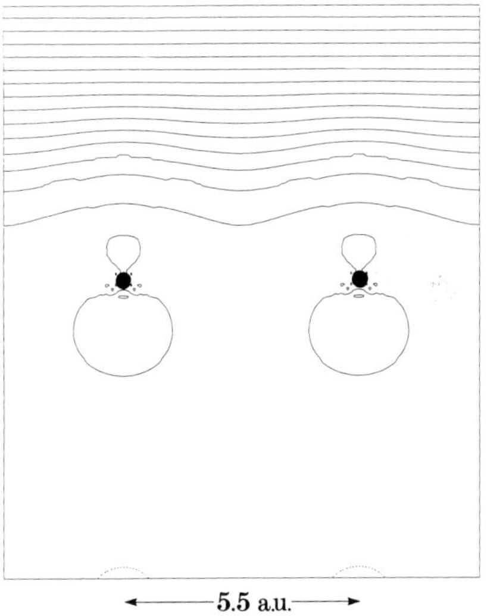

Figure 3

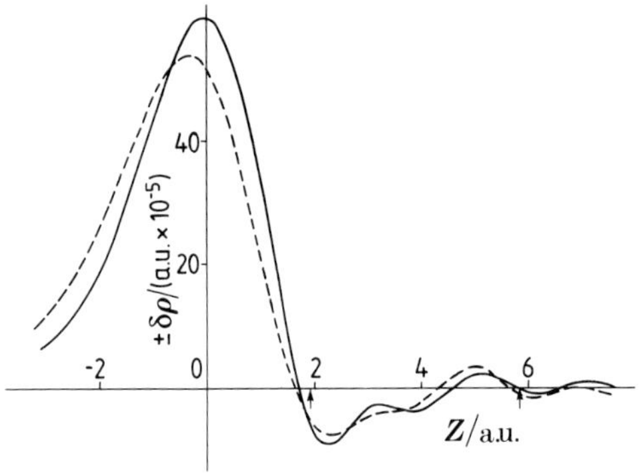

Figure 2. Change in potential at Ag (001) with field $\mathscr{E} = 0.01$ a.u.

Figure 3. Planar-averaged screening charge at Ag (001) for fields $\mathscr{E} = +0.02$ a.u. (solid line), and $\mathscr{E} = -0.02$ a.u. (dashed line), as a function of distance from the geometrical surface. Arrows mark the surface and subsurface atomic planes.

Our calculated image plane can be compared with the results of inverse photoemission experiments. By treating z_0 as a parameter in a model surface potential and using this to fit the experimentally observed energies of the Rydberg surface states, Smith *et al.* (1989) conclude that the image plane at Ag (001) lies at $+0.18$ a.u., on the *solid* side of the surface. Our result is in marginally better agreement with experiment than the jellium result (Weber & Liebsch 1987*a*), $z_0 = -1.35$ a.u., found using jellium of density $r_s = 3$ a.u., which is frequently used to model the response of Ag surfaces to external fields. However, there does seem to be a real discrepancy between our theory and experiment – what might the reasons for this be? From (8), nonlinear effects – due to the size of the electric field produced by the electron in the surface state – are not big enough to account for the discrepancy. A fundamental problem is that the calculation is carried out within the local density approximation, which does not correctly describe the image potential in the Schrödinger equation. Several authors have tried to improve on this for jellium: Ossicini *et al.* (1987) used the non-local exchange-correlation potential of the weighted density approximation in their determination of the screening charge, and Serena *et al.* (1988) used an interpolation scheme between the asymptotic form of the image potential and the local density potential nearer the surface. For jellium with $r_s = 3$ a.u. these corrections push the image plane 0.1 a.u. closer to the surface. However, the real problem is that our calculation gives the response of the surface to a static external charge, whereas an electron in a Rydberg surface state is a dynamic object. We ought, in fact, to be calculating the asymptotic form of the *self-energy* (Inkson 1971).

The position of the image plane at Al surfaces has been calculated by several authors, and it provides a sensible system for comparison with the jellium results of Lang & Kohn (1973): for $r_s = 2$ a.u., $z_0 = -1.6$ a.u. We find that the centre of gravity of the screening charge for Al (001) in an electric field of $+0.01$ a.u. lies at $z_0 =$

-1.1 a.u., significantly closer in than for jellium. In their pseudopotential slab calculation, in which the z-dependent planar-averaged potential was used, Serena *et al.* (1988) find the image plane 0.26 a.u. closer in to the surface than the jellium result. Another comparison is with the work of Finnis (1990), who puts the image plane for Al (111) at -0.87 a.u. Although there are discrepancies between the different calculations, it appears that the image plane lies closer to the surface when atoms are included than in the jellium calculation. The reason for this is that the electrons feel a more attractive electrostatic potential than in jellium, where the bulk electrostatic potential from the electrons and compensating positive background is zero. The more attractive potential reduces the spilling-out of the electrons at the surface, hence pulls in the image plane.

The field-dependence of the screening charge profile gives a second-order current normal to the surface, and the intensity of the second harmonic reflected from the surface in a low-frequency external field is proportional to the square of the coefficient of $\mathscr{E}$ in (8) (Weber & Liebsch 1987*a*). This is of considerable interest nowadays, because second harmonic generation is very surface-sensitive, and can be greatly enhanced by small quantities of alkali adsorbates for example (Weber & Liebsch 1987*b*). Our results show that the screening charge at the Ag (001) surface is much stiffer than jellium calculations indicate: our value of 8.83 for the coefficient of $\mathscr{E}$ in (8) is to be compared with the value of 30 a.u. found for jellium with $r_s =$ 3 a.u. In other words, we predict a second harmonic intensity a factor of 12 smaller than for a jellium model of Ag (001). This is in the right direction to obtain agreement with experiment, as Guyot-Sionnest *et al.* (1990) have shown in a recent study of a Ag electrode–electrolyte interface. A possible explanation for the greater stiffness of the screening charge in the Ag (001) calculation is that the ion cores seem relatively impenetrable (figure 1), thereby pinning the screening charge.

4. Surface effective charges

The force normal to the surface on atoms i in an external electric field $\mathscr{E}$ can be used to define their effective charge:

$$F_i = q_i^* \mathscr{E}. \tag{9}$$

From Hellmann–Feynman, F_i is the fully screened field at nucleus i times the nuclear charge, so q_i^* is a measure of the effectiveness of the screening; from the results for the screened potential shown in figure 2 we would expect q_i^* to be normally rather small. As elementary electrostatics tell us that the total force on the surface is proportional to $\mathscr{E}^2$ and the force linear in $\mathscr{E}$ vanishes, it follows that:

$$\sum_i q_i^* = 0, \tag{10}$$

a result discussed many years ago by Trullinger & Cunningham (1973).

The surface effective charge gives the variation in work-function ϕ with atomic displacements (Hamann 1987). Let us consider the variation $\delta\phi$ when all atoms of type i (this will be a layer of atoms parallel to the surface) are displaced by δz normal to the surface. As the work-function is given by the energy change in removing an electron through the surface to infinity, we can write $\delta\phi$ as:

$$\delta\phi = (\partial E_{N-1}/\partial z_i - \partial E_N/\partial z_i)\,\delta z_i, \tag{11}$$

where E_N is the energy of the charge-neutral system with N electrons, and E_{N-1} is the

Table 1

external field (a.u.)	field at H (a.u.)
0	−0.004431
0.005	−0.004327
0.01	−0.004358
0.02	−0.004261

energy of the charged system with one electron removed. Now $\partial E_N/\partial z_i$ is zero in equilibrium, and $\partial E_{N-1}/\partial z_i$ is (minus) the force on the $\mathcal{N}$ atoms of type i in the charged system. But in this charged system (we assume it is metallic), the charge deficit is at the surface, so there is an external field given by:

$$\mathscr{E} = -4\pi|e|/\mathcal{N}A, \tag{12}$$

where A is the area of the surface unit mesh. Hence we obtain the connection between the work-function variation and the effective charge:

$$\partial\phi/\partial z_i = 4\pi q_i^*/A. \tag{13}$$

In this expression the convention is that the electronic charge is negative, and z is directed into the surface. As the variation in work-function with atomic displacement comes entirely from the change in the surface dipole layer, this expression has just the form we would expect from moving charges q_i^* rigidly, and it is analogous to the standard theory of effective charges in ionic crystals.

With a surface phonon, the variation of the surface dipole over the surface via (13) sets up long-range electric fields which scatter electrons in EELS experiments (Ibach 1971). An interesting case is a monolayer of H on Rh (001), where EELS experiments show evidence of dipole-active modes, corresponding to a finite effective charge on the H atoms (Richter & Ho 1987). However, Hamann & Feibelman (1988) found in a slab calculation, with H adsorbed on either side of a 3-layer Rh (001) film, that the work-function is remarkably constant as a function of the H–Rh interlayer spacing, suggesting an effective charge of zero! To try to resolve this, we have carried out an embedded calculation for Rh (001) (1×1)–H, treating explicitly the H and the top layer of Rh, and embedding this on to bulk Rh (Miller & Inglesfield 1991). As we have seen, embedding eliminates the finite size effects associated with the slab. From the change in work-function around the equilibrium interlayer spacing (1.1 a.u. with the H atoms in four-fold hollow sites), we find that the effective charge on the H atoms is $+0.009|e|$. Next, we determined the effective charge from (9), applying an electric field to the surface and finding directly the screened field at the H nucleus. Our results, presented in table 1, show considerable scatter, and in the case of zero external field there is actually a field at the H due to the fact that one layer of Rh is not enough to do the equilibrium energetics accurately. Nevertheless, a straight line of slope $+0.009$ – corresponding to the effective charge of $+0.009|e|$ – provides a reasonable fit.

This effective charge for H atoms adsorbed on Rh (001) is much smaller than is found on other substrates: $q_{\mathrm{H}}^* = +0.054|e|$ for H/Pd (111) and $-0.032|e|$ for H/W (001) (Hamann 1987), and it remains to be seen whether it is compatible with the EELS results. This substrate dependence is not just a question of electronegativity, because Rh and Pd are the same on the Pauling scale; we do find a structure dependence to the effective charge (Miller & Inglesfield 1990).

5. Effective charges and surface instability

A large effective charge can lead to a surface instability. This was first shown by Trullinger & Cunningham (1973) using a pair force expression for the dynamical matrix, including long-range Coulomb forces between the effective charges, but we shall use a more general, macroscopic argument.

We consider a normal displacement of atom i, as in (13), but with an amplitude varying sinusoidally over the surface with wavevector $\boldsymbol{K}$:

$$\delta z_i(\boldsymbol{R}) = a_i \sin(\boldsymbol{K}\cdot\boldsymbol{R}). \tag{14}$$

From (13), this sets up a spatially varying surface barrier potential, and because the bulk Fermi energy is constant, the electrostatic potential outside the surface varies as:

$$V(\boldsymbol{R}, z) = -(4\pi/A)\, q_i^* \, a_i \sin(\boldsymbol{K}\cdot\boldsymbol{R}) \exp(Kz). \tag{15}$$

The normal component of the corresponding 'patch' electric field near the surface is:

$$\mathscr{E} = (4\pi/A)\, q_i^* \, K a_i \sin(\boldsymbol{K}\cdot\boldsymbol{R}), \tag{16}$$

so from (9) the force on the atoms is:

$$F = (4\pi/A)\,(q_i^*)^2 K a_i \sin(\boldsymbol{K}\cdot\boldsymbol{R}), \tag{17}$$

proportional to the displacements of the atoms, and countering the local restoring forces. If the restoring force on atoms i in the long-wavelength limit is α_i, we see that the surface is unstable for displacement wavelengths shorter than

$$\lambda = 8\pi^2 (q_i^*)^2 / \alpha_i A. \tag{18}$$

This suggests that any surface with non-vanishing q_i^* will be unstable for sufficiently short wavelengths, but of course the argument breaks down once λ is of the order of an interatomic spacing.

It was suggested by Trullinger & Cunningham (1973) that an effective-charge-driven instability might be responsible for semiconductor surface reconstructions. However, it is the dangling bonds which drive the reconstructions in these cases. But one candidate for such a reconstruction is O adsorbed on Cu (001), the $(\sqrt{2}\times\sqrt{2})$ $R45°$ structure, with O atoms in four-fold hollow sites suggested by analogy with O/Ni (001), is unstable (Asensio *et al.* 1990). From our calculations (Colbourn & Inglesfield 1991), the effective charge of the O atoms in these ideal sites is $-0.9|e|$. Taking the force constant from EELS measurements (Wuttig *et al.* 1989) as 0.072 a.u. we find that the surface is unstable for wavelengths smaller than 19 a.u. But this is considerably greater than the O–O interatomic spacing, so the surface must be unstable in the ideal structure.

I have had very useful discussions on effective charges with Wolfram Miller and Elizabeth Colbourn. The work has been supported by FOM (Stichting voor Fundamenteel Onderzoek der Materie).

References

Aers, G. C. & Inglesfield, J. E. 1989 Electric fields and Ag (001) surface electronic structure. *Surf. Sci.* **217**, 367–383.

Appelbaum, J. A. & Hamann, D. R. 1972 Self-consistent electronic structure of solid surfaces. *Phys. Rev.* B**6**, 2166–2177.

Appelbaum, J. A. & Hamann, D. R. 1973 Surface states and surface bonds of Si (111). *Phys. Rev. Lett.* **31**, 106–109.

Asensio, M. C., Ashwin, M. J., Kilcoyne, A. L. D., Woodruff, D. P., Robinson, A. W., Lindner, Th.,

Somers, J. S., Ricken, D. E. & Bradshaw, A. W. 1990 The structure of oxygen adsorption phases on Cu (100). *Surf. Sci.* **236**, 1–14.

Colbourn, E. A. & Inglesfield, J. E. 1991 Surface instability and electronic structure of O on Cu (001). (In preparation.)

Finnis, M. W. 1990 The interaction of a point charge with an aluminium (111) surface. *Surf. Sci.* (In the press.)

Fu, C. L. & Ho, K. M. 1989 External-charge-induced surface reconstruction on Ag (110). *Phys. Rev. Lett.* **63**, 1617–1620.

Guyot-Sionnest, P., Tadjeddine, A. & Liebsch, A. 1990 Electronic distribution and non-linear optical response at the metal-electrolyte interface. *Phys. Rev. Lett.* **64**, 1678–1681.

Hamann, D. R. 1987 Hydrogen vibrations at transition metal surfaces. *J. Elect. Spectroscopy* **44**, 1–16.

Hamann, D. R. & Feibelman, P. J. 1988 Anharmonic vibrational modes of chemisorbed H on the Rh (001) surface. *Phys. Rev.* B **37**, 3847–3855.

Heine, V. 1963 On the general theory of surface states and scattering of electrons in solids. *Proc. Phys. Soc.* **81**, 300–310.

Heine, V. 1970 The pseudopotential concept. *Solid St. Phys.* **24**, 1–36.

Ibach, H. 1971 Surface vibrations of silicon detected by low-energy electron spectroscopy. *Phys. Rev. Lett.* **27**, 253–256.

Inglesfield, J. E. 1981 A method of embedding. *J. Phys.* C **14**, 3795–3806.

Inglesfield, J. E. 1990 Electric fields and surface electronic structure. *Vacuum* **41**, 543–546.

Inglesfield, J. E. & Benesh, G. A. 1988 Surface electronic structure: embedded self-consistent calculations. *Phys. Rev.* B **37**, 6682–6700.

Inkson, J. C. 1971 The electron–electron interaction near an interface. *Surf. Sci.* **28**, 69–76.

Kolb, D. M., Boeck, W., Ho, K.-M. & Liu, S.-H. 1981 Observation of surface states on Ag (100) by infrared and visible electroreflectance spectroscopy. *Phys. Rev. Lett.* **47**, 1921–1924.

Krakauer, H., Posternak, M. & Freeman, A. J. 1979 Linearized augmented plane-wave method for the electronic band structure of thin films. *Phys. Rev.* B **19**, 1706–1719.

Lang, N. D. & Kohn, W. 1970 Theory of metal surfaces: charge density and surface energy. *Phys. Rev.* B **1**, 4555–4568.

Lang, N. D. & Kohn, W. 1973 Theory of metal surfaces: induced surface charge and image potential. *Phys. Rev.* B **7**, 3541–3550.

Liebsch, A. & Schaich, W. L. 1989 Second-harmonic generation at simple metal surfaces. *Phys. Rev.* B **40**, 5401–5410.

Melmeed, A. J., Tung, R. T., Graham, W. R. & Smith, G. D. W. 1979 Evidence for reconstructed (001) tungsten obtained by field-ion microscopy. *Phys. Rev. Lett.* **43**, 1521–1524.

Miller, W. & Inglesfield, J. E. 1991 Effective charge of H on Rh (001). (In preparation.)

Ossicini, S., Finocchi, F. & Bertoni, C. M. 1987 Electron density profiles at charged metal surfaces in the weighted density approximation. *Surf. Sci.* **189/190**, 776–781.

Richter, L. J. & Ho, W. 1987 Vibrational modes of hydrogen adsorbed on Rh (100) and their relevance to desorption kinetics. *J. Vac. Sci. Technol.* A **5**, 453–454.

Serena, P. A., Soler, J. M. & Garcia, N. 1988 Work function and image-plane position of metal surfaces. *Phys. Rev.* B **37**, 8701–8706.

Shaw, R. W. Jr & Harrison, W. A. 1967 Reformulation of the screened Heine–Abarenkov model potential. *Phys. Rev.* **163**, 604–611.

Smith, N. V., Chen, C. T. & Weinert, M. 1989 Distance of the image plane from metal surfaces. *Phys. Rev.* B **40**, 7565–7573.

Trullinger, S. E. & Cunningham, S. L. 1973 Soft-mode theory of surface reconstructions. *Phys. Rev. Lett.* **30**, 913–916.

Weber, M. G. & Liebsch, A. 1987*a* Density-functional approach to second-harmonic generation at metal surfaces. *Phys. Rev.* **35**, 7411–7416.

Weber, M. G. & Liebsch, A. 1987*b* Theory of second-harmonic generation by metal overlayers. *Phys. Rev.* B **36**, 6411–6414.

Wuttig, M., Franchy, R. & Ibach, H. 1989 Oxygen on Cu (100) – a case of an adsorbate induced reconstruction. *Surf. Sci.* **213**, 103–136.

Discussion

D. WEAIRE (*Dublin, Republic of Ireland*). The embedding method really comes into its own for SHG because a finite slab must have two surfaces, whose contributions can actually cancel in a naive calculation! How does Professor Inglesfield calculate the SHG from the single surface in his formation?

J. E. INGLESFIELD. This is calculated from the field dependence of the shape of screening charge, in the response of the surface to an external field. At the moment it is only possible to calculate static response when a real surface is treated, so the calculation is really only applicable in the low-frequency limit; it cannot treat spectroscopic aspects of SHG.

O. K. ANDERSEN (*Stuttgart, F.R.G.*). At the meeting we have heard a lot about pseudopotentials and local density functional potentials. The advantage of such potentials is that they are energy independent, local in r-space, weak, etc. Professor Inglesfield's surface electronic structure method uses a so-called embedding potential to describe the semi-infinite bulk part of the system, but in his talk he did not tell us about the properties and the advantages of using this potential. Is it not strongly energy dependent, non-local and so on? How does Professor Inglesfield's method compare with Green's function techniques (e.g. matching – or Dyson's – methods)?

J. E. INGLESFIELD. The embedding potential is non-local, energy dependent and complex; inevitably non-local because it is describing the scattering properties of the whole substrate, energy dependent because this is connected with the normalization of the wavefunctions in the substrate, and complex to smear out the discrete states of the finite surface region into the continuum associated with the semi-infinite solid.

We have compared embedding results with Pollmann's work on Si (001) using a Dyson's equation approach and found good agreement; embedding offers greater freedom in the choice of basis set, but this is not always an advantage. It has an advantage over matching Green function methods once the embedding potential has been determined, because embedding methods can use fairly standard band structure technology.

U. GERHARDT (*Frankfurt, F.R.G.*). The system oxygen on Cu (001) is a very tricky one. We actually found out by LEED investigations that the adsorption site on a very smooth Cu (001) surface, i.e. one without surface steps, is the bridge site. The fourfold hollow site only shows up if such steps are present. This might, however, be compatible with the surface instability mentioned after all.

J. E. INGLESFIELD. This is a very interesting result; I cannot deduce, from my surface instability argument, what the surface will go to. Clearly O on Cu (001) is a very difficult and remarkable surface.

V. HEINE (*Cambridge, U.K.*). The essence of jellium to my mind is that the potential in the bulk is a constant. The question is what value to take for that constant. If we start from a pseudopotential picture of a real metal, then one should take the mean Hartree pseudopotential. Unfortunately in the jellium calculations done by most people, the Hartree potential is effectively set equal to zero, which seems to me physically quite unrealistic as a model of real metal. In both cases the (attractive) exchange and correlation potential is added: in my approach one must be careful not to count that twice. Of course in the infinite bulk solid it makes no difference which one does: indeed there is no way of defining a zero of potential. But at a surface the difference between the two jellium models is quite significant, and I am therefore not surprised by the difference in the second-order response to a surface electric field that was mentioned.

Bonding at surfaces

By J. B. Pendry
The Blackett Laboratory, Imperial College of Science, Technology and Medicine, London SW7 2BZ, U.K.

Surface crystallography is a rapidly growing subject in terms of both its scope and capabilities. Some surface structures are examined in detail and their interesting features stressed. We go on to examine new developments in the field, and conclude with a review of possibilities for studying thermal motion at surfaces of both harmonic and anharmonic nature, and of thermal diffusion across the surface.

1. Overview of structures

Surface crystallography has come of age and the atomic arrangement of a surface can now be reliably determined for all but the most complex situations. Something like 500 surface structures are reported in the Surface Crystallographic Information Service catalogue (MacLaren *et al.* 1987) and with recent additions to the literature close to 1000 structures have now been determined, most of these by low-energy electron diffraction, but with a wide variety of other techniques contributing: surface X-ray diffraction, Rutherford back scattering, surface extended X-ray absorption fine structure, to name the major contenders. Enough is now known that some patterns are emerging which are unique to surface crystallography, and relevant to a host of phenomena from semiconductor growth to catalysis.

The plan of this article is to review a few of the known surface structures, to outline how LEED functions (the major technique of surface determination) and then to speculate about future directions which surface crystallography may explore.

Figure 1 shows the structure formed when Ni is deposited on a Si (111) surface, as might be the case in making a Schottky diode. The structure, determined by Van Loenen *et al.* (1985), clearly shows that the interface consists of a chemical compound, $NiSi_2$, not of pure metal. This simple structural fact explains the many anomalies in diode performance, which early theories predicted on the basis of the barrier height given by a pure metal–semiconductor interface.

Figure 2 is another overlayer structure: that of C adsorbed on Ni (100) (Onuferko *et al.* 1979). For obvious reasons this has been christened 'the clock structure'. It illustrates a very common situation in adsorbate systems, namely that the substrate is not simply a passive template but itself responds to the presence of an adsorbate. Sometimes the response is a simple one with substrate inter layer spacing relaxations, sometimes it can be extremely complex as in the case of N adsorbed on W (100). Here we see a beautifully symmetrical structure. The eye is naturally drawn to the alternate rotation of four Ni atoms about the carbon sites, but the driving force is almost certainly to be found on the sites holding no C atoms. Here four substrate Ni atoms are rearranging themselves from a square structure to an hexagonal structure, packing closer together in the process, and releasing more space around the C sites.

The C on Ni (100) structure is a example of an extremely important class of

Figure 1

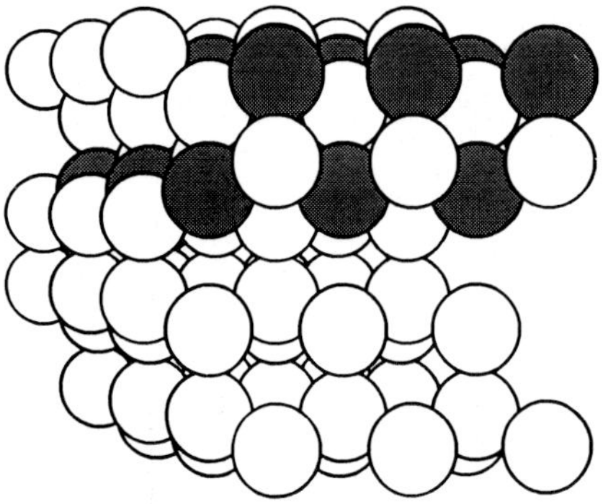

Figure 2

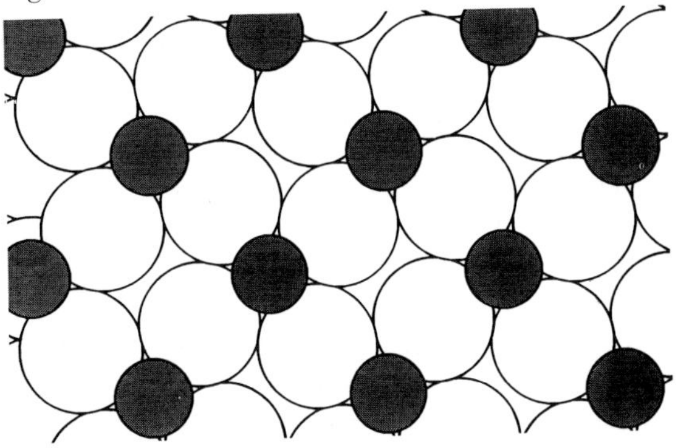

Figure 1. The $NiSi_2$ (111)–Si (111) interface as determined by Rutherford back scattering (Van Loenen *et al.* 1985). Si atoms are shown as open circles. This structure is formed when Ni is deposited on Si (111) at 300 K followed by annealing.

Figure 2. The 'clock structure' of P(2 × 2)C/Ni (100) as determined by Onuferko *et al.* (1979). Note the distortion of the top layer of Ni: the clock-like rotation about the C sites, and the hexagonal distortion about the empty sites.

systems: carbidic carbon adsorbed on transition metal surfaces. Billions of dollars worth of organic chemicals are formed in processes involving similar structures. We are taking the first steps in producing a detailed understanding of these systems.

Figure 3 shows a structure which is also relevant to the catalysis industry, that of CO adsorbed on Cu (100). The major catalyst for methanol formation from CO is Cu. This structure, determined by Andersson & Pendry (1980), is typical of the molecular phase of interaction with surfaces. The CO molecule is relatively weakly bound, sits well clear of the substrate atoms in contrast to C on Ni. Many molecules have such weak interactions with substrates, preserving their vibrational frequencies except for small shifts. This means that they are highly mobile at the surface, and can easily desorb, a most desirable characteristic in a reaction product.

The structure shown in figure 4 (Van Hove *et al.* 1986) is a tour de force representing the state of the art for surface structure determination. It shows two adsorbates, CO and C_6H_6, adsorbed on a Rh (111) surface. Several interesting points are raised here. It is observed that neither CO nor C_6H_6 form separate structures on this surface. Only when both are present does a well-ordered surface structure form. The reason appears to be that charge transfer occurs from C_6H_6 to CO and electrostatic forces stabilize the resulting surface ionic compound. Similar structures are seen when CO is coadsorbed with alkali metals. Look carefully at the C_6H_6 and you will see that it is no longer six-fold symmetric. The C atoms now cluster in three acetylene pairs on top of each of the three Rh atoms. This distortion of the six-fold ring is about the largest observed without the molecule fragmenting. Could we have caught the molecule on the verge of a reaction?

All the systems above form ordered structures, and conventional methods require this before experiments can be analysed to give the atomic coordinates. Two methods are capable of side stepping this restriction: surface EXAFS and diffuse LEED. The latter method has been applied to disordered O adsorbed on W (100) by Rous *et al.* (1986). The structure is shown in figure 5*a* and compared with the structure of the reconstructed clean W (100) surface in figure 5*b* (Barker *et al.* 1978). The clean

Figure 3

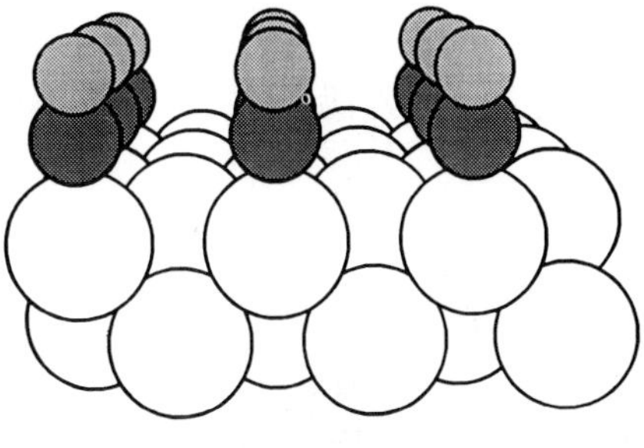

Figure 4

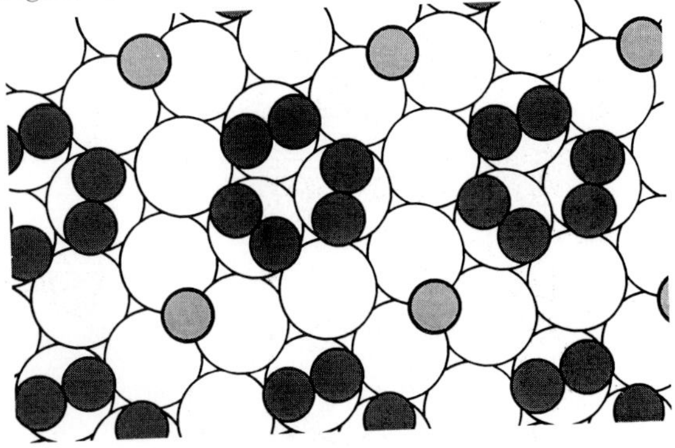

Figure 3. C(2 × 2)CO/Cu (100) determined by Andersson & Pendry (1980). The open circles represent Cu, the black circles C, and the half-shaded circles O. Contrast the location of C atoms in this structure at on-top sites, with that in figure 2 where isolated C atoms bury into the four-fold hollows.

Figure 4. A complex surface structure determined by Van Hove *et al.* (1986) containing a mutually stabilizing arrangement of CO and C_6H_6 molecules on the Rh (111) surface. H atoms are not shown, and the C in the CO molecules is hidden by the terminal O. Note the restructuring of C_6H_6 induced by the underlying Rh atoms.

(*a*) (*b*)

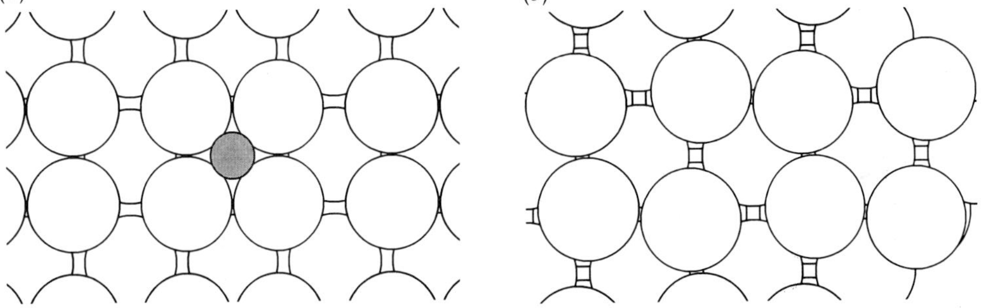

Figure 5. O adsorbs on W (100) in a disordered state at low temperatures, but its structure can still be determined by diffuse LEED (Rous *et al.* 1986). Note how in (*a*) the top layer of W atoms reconstructs around the O, in contrast to the clean surface reconstruction shown in (*b*), as determined by Barker *et al.* (1978).

surface forms zig-zag chains running across the surface in an effort to acquire some in-plane coordination for the surface W atoms. O acts as a nucleus around which these chains can wrap themselves. In this way the W atoms keep two in-plane W neighbours, but in addition acquire some energy from bonding to O. Once again an adsorbate has radically changed the substrate structure. It is almost as if each adsorbate atom is elastically screened by substrate movement, just as an adsorbed ion is electronically screened by substrate electrons. How much energy is involved, and whether screening can play a role in smoothing the course of chemical reactions remains to be investigated.

These structures illustrate the uniqueness of the surface environment. Surfaces are not simply a corner of the bulk solid. In the bulk atoms are strongly confined by the geometry. The consequence of weak bonds tends to be melting or evaporation, but at the surface it is only necessary to confine the atoms normal to the surface. The

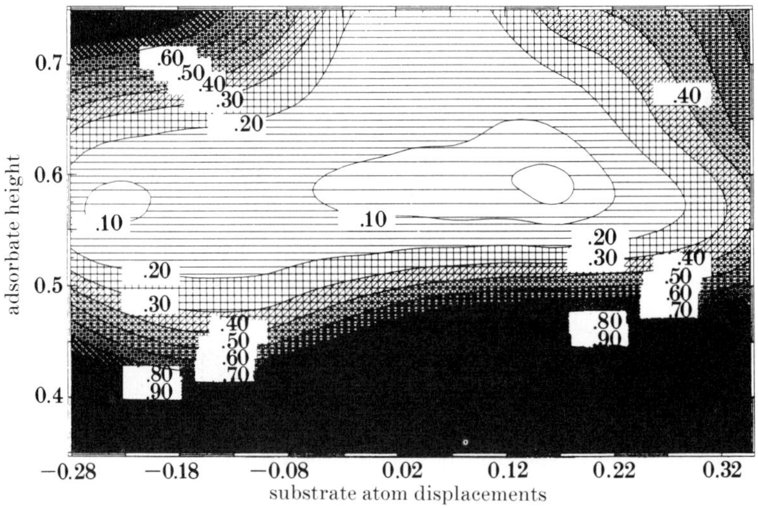

Figure 6. A typical R-factor map. This one is for the O/W (100) system shown in figure 5*a*. The R-factor summarizes agreement between theory and experiment for a trial structure: zero for perfect agreement, 1 for perfect disagreement. The trial variables are: the height of the O above the surface, and the magnitude of the top layer W displacements. A clear minimum is seen indicating the correct structure.

parallel motion can be harmonic, anharmonic, or gas like. Thus the surface is not simply a two-dimensional crystal, but is more analogous to a two-dimensional world, encompassing gas, liquid as well as solid phases.

2. Developments in methods of structural analysis

Traditional surface structure determination has been made by trial and error analysis of experimental data (see, for example, Pendry 1974; Van Hove & Tong 1979; Heinz & Mueller 1982). For example, in the case of low-energy electron diffraction, intensities of reflected beams are collected as functions of incident energy: perhaps six different beams over an energy range 50–250 eV. A large quantity of data is important because a great deal of surface structure information is encoded in the data, and a large data-set is essential before all this information can be reliably extracted. Because of the well-known complications of multiple scattering of electrons, the connection between structure and diffraction data is not trivial, and extraction of information is done by postulating a trial structure, calculating the spectra implied by that structure, and comparing to experiment to test how close a guess we have made. It is usual to sum up the theory–experiment agreement for a trial in a single number called the R-factor (Zannazzi & Jona 1977; Pendry 1980) which is defined to be 0 for perfect agreement, and 1 for perfect disagreement. Figure 6 shows an example for the disordered O on W (100) structure taken from Rous *et al.* (1986). The R-factor is plotted as a function of the trial vertical height of the O and of the size of the W reconstruction. There is a clearly identifiable minimum which represents the true structure. Complications arise when data are of poor quality, or of small quantity. Then several trial structures may fit the data equally well and confusion has sometimes arisen in the past for these reasons.

Another major difficulty arises when there are many unknown coordinates in a surface structure. Suppose that we wish to try 10 possible values for each of six

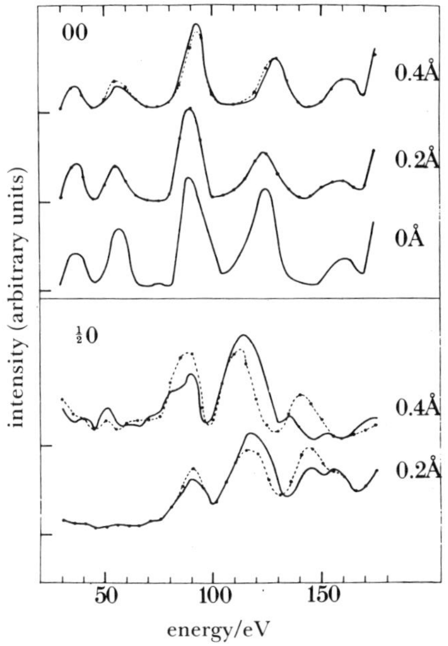

Figure 7. A test of the accuracy of tensor LEED taken from Rous *et al.* (1986): full dynamical (——) and tensor LEED (– – –) calculations for an artificial P(2 × 2) reconstruction of the Ni (100) surface in which every fourth atom is raised by 0, 0.2, 0.4 Å in the three curves shown. The method appears to be accurate for displacements as large as 0.4 Å.

coordinates. The number of trials is 10^6 and grows exponentially with the number of coordinates to be determined. The consequence is that for complex structures there is nearly always an inadequate search of parameter space. To combat this difficulty some new ideas were introduced into the theoretical calculations which are the most time-consuming step in data analysis (Rous *et al.* 1986; Rous & Pendry 1989*a*, *b*).

The idea is to introduce a reference structure which is a best guess at the true structure. One calculation of LEED intensities is made for the reference structure and closely related structures are then handled by first-order perturbation theory. It is easy to show that the change in diffracted amplitude of the **g**th beam can be written,

$$\delta A(\boldsymbol{g}) \approx \Sigma_j \boldsymbol{M}(\boldsymbol{g}, j)\, S(j, \delta \boldsymbol{R}), \qquad (1)$$

where S is purely a geometrical function of the displacements, $\delta\boldsymbol{R}$, away from the reference structure; the tensor $\boldsymbol{M}$ is calculated once and for all for a given reference structure. Obviously matrix multiplication is made very rapidly and many trials can be made at little cost. In many cases the whole procedure is speeded by a factor of 1000 and the analysis can be made on a personal computer for relatively simple systems. The name given to the methodology is 'tensor LEED'.

In figure 7 we show a trial of the new method for a hypothetical structure taken from Rous *et al.* (1986). The trial structure is an unreconstructed Ni (100) surface. The perturbation is a reconstruction of this surface formed by raising every fourth atom in the surface in a P (2 × 2) array. The integer order beams, present also in the unreconstructed case, show excellent agreement with the exact calculation for displacements of up to 0.4 Å†, and even the fractional order beams which are unique to the reconstructed surface show acceptable agreement out to this magnitude of

† 1 Å = 10^{-10} m.

displacement. Displacements of this magnitude encompass most adsorbate induced surface reconstructions. When the atoms concerned are very light ones such as C or O, then the method works even better because of their weak scattering nature and light atoms can in fact be displaced anywhere on the surface and still be treated by perturbation theory.

3. Direct methods

Tensor LEED is a very effective way of analysing more complex structures, based on very straightforward concepts. However, the next step examines the new concept introduced and makes serious analysis of the possibilities which it holds (Pendry & Heinz 1990).

Suppose that, as before, we have defined a reference structure, and that the true structure is displaced from the reference by a small distance, $\delta \boldsymbol{R}$. Then the potential of the surface changes by

$$\delta V \approx \delta \boldsymbol{R} \cdot \nabla V(\boldsymbol{r}-\boldsymbol{R}). \tag{2}$$

Provided that $\delta \boldsymbol{R}$ is sufficiently small we can use first-order perturbation theory to calculate,

$$\begin{aligned} \delta A(\boldsymbol{g}) &\approx \langle \boldsymbol{k}+\boldsymbol{g}; \text{out} | \delta \boldsymbol{R} \cdot \nabla V | \boldsymbol{k}; \text{in} \rangle \\ &= \Sigma_j \boldsymbol{M}'(\boldsymbol{g}, j)\, \delta \boldsymbol{R}_j. \end{aligned} \tag{3}$$

Hence if we take this equation seriously, and if we can measure enough values of $\delta A(\boldsymbol{g})$, then equation (3) can be inverted to give,

$$\delta \boldsymbol{R}_j \approx \Sigma_g \boldsymbol{M}'^{-1}(j, \boldsymbol{g})\, \delta A(\boldsymbol{g}). \tag{4}$$

There is the question of how to measure $\delta A(\boldsymbol{g})$, but we can obtain this easily for small $\delta A(\boldsymbol{g})$,

$$\begin{aligned} |A(\boldsymbol{g})+\delta A(\boldsymbol{g})| &= |A(\boldsymbol{g})|^2 + (A^*(\boldsymbol{g})\, \delta A(\boldsymbol{g}) + A(\boldsymbol{g})\, \delta A^*(\boldsymbol{g})) + |\delta A(\boldsymbol{g})|^2 \\ &\approx |A(\boldsymbol{g})|^2 + (A^*(\boldsymbol{g})\, \delta A(\boldsymbol{g}) + A(\boldsymbol{g})\, \delta A^*(\boldsymbol{g})), \end{aligned} \tag{5}$$

and hence with a knowledge of $A(\boldsymbol{g})$ we have the information we require.

In principle, when $\delta \boldsymbol{R}_j$ is small, we can solve for the structure by a direct inversion procedure without having to use trial and error at all. Obviously if this were possible all the difficulties of having to make good guesses and to make huge numbers of trial calculations would be circumvented. The problem is that matters are not quite so simple: equation (2) is a very bad approximation for all but very small values of $\delta \boldsymbol{R}_j$, essentially because the wavelength of 150 eV electrons is of the order of 1 Å. We can get around the problem as follows by using the exact atomic t-matrix,

$$\begin{aligned} \delta A(\boldsymbol{g}) &\approx \langle \boldsymbol{k}+\boldsymbol{g}; \text{out} | t(\boldsymbol{R}+\delta \boldsymbol{R}) - t(\boldsymbol{R}) | \boldsymbol{k}; \text{in} \rangle \\ &= \Sigma_n \boldsymbol{M}(\boldsymbol{g}, n)\, \delta \boldsymbol{R}^n, \end{aligned} \tag{6}$$

that is, we have made a polynomial expansion in powers of $\delta \boldsymbol{R}$. The interpretation of (6) is clearer if we recognize that we are determining, not the location of a single atom, but an average location of many atoms at the surface, relative to the reference structure. Thus after inversion we find,

$$\langle \delta \boldsymbol{R}^n \rangle \approx \Sigma_g \boldsymbol{M}^{-1}(n, \boldsymbol{g}) \langle \delta A(\boldsymbol{g}) \rangle, \tag{7}$$

where the brackets $\langle\ \rangle$ denote an average. Now we see that our first shot at theory though equation (4) merely gave the average location of atoms. Although we now

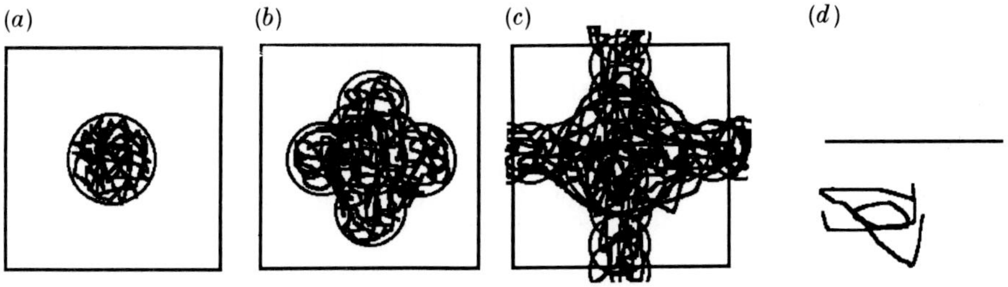

Figure 8. Diffusion of an adsorbed atom around the unit cell of a surface. Note how the motion can be expected to change in character as the amplitude of motion increases, first becoming anharmonic, and then diffusive, as the atom breaks loose from its site. (*a*) Low temperature, spherical distribution, harmonic vibrations; (*b*) intermediate temperature, non-spherical distribution, anharmonic vibrations; (*c*) high temperature, distribution overlaps cell sides, surface diffusion; (*d*) unit surface, adsorbate trajectory.

pay the price of a more complex theory, the reward is that we retrieve not just the atom positions but the moments of their distribution, $P(\delta \boldsymbol{R})$. If we know,

$$\langle \delta \boldsymbol{R}^n \rangle = \int \delta \boldsymbol{R}^n P(\delta \boldsymbol{R}) \, \mathrm{d}\delta \boldsymbol{R}, \tag{8}$$

then we can reconstruct $P(\delta \boldsymbol{R})$ by standard methods.

Developing the direct methods has given us more than we hoped for. Diffraction measurements can now give information not just about static locations of atoms, but also about their distribution.

4. Thermal motion of atoms at surfaces

We have already discussed how surfaces are differentiated from the bulk of a solid by the freedom they allow to adsorbed atoms in the plane parallel to the surface. This unique property is almost entirely without an experimental probe. Currently surface crystallography studies averaged atom positions. Surface vibrations are accessible through probes which lose energy to the phonons, but study mainly high-frequency harmonic vibrations of strong bonds which anchor the atom or molecule firmly in place. Much more interesting are the very-low-frequency vibrations of molecules about to dissociate, or the diffusive motion of a molecule moving from one reaction site to the next. In figure 8 we show examples of various types of thermal motion at a surface.

Direct methods hold the potential for extending experimental surface studies to the fascinating and vital area of low-frequency vibration and diffusion.

The idea is a simple one. If we can assume that the atoms and molecules at a surface are disordered but in thermal equilibrium, then if the motion is classical and not quantum (a complication with H) we can apply the Boltzmann formula,

$$P(\delta \boldsymbol{R}) = \text{const.} \times \exp\left(-V_{\mathrm{A}}(\delta \boldsymbol{R})/k_{\mathrm{B}} T\right), \tag{9}$$

where $V_{\mathrm{A}}(\delta \boldsymbol{R})$ is the potential in which the atom migrates across the surface. Our prescription for finding $V_{\mathrm{A}}(\delta \boldsymbol{R})$ is to apply direct methods to the surface at several different temperatures and extract $P(\delta \boldsymbol{R})$ at each temperature. This will enable us to infer $V_{\mathrm{A}}(\delta \boldsymbol{R})$. Several temperatures must be used because at any given temperature (9) only gives good sensitivity when $V_{\mathrm{A}}(\delta \boldsymbol{R})$ is of the order,

$$V_{\mathrm{A}}(\delta \boldsymbol{R}) \approx k_{\mathrm{B}} T. \tag{10}$$

Once in possession of $V_A(\delta \boldsymbol{R})$ some elementary molecular dynamics simulation will give the harmonic and anharmonic vibrations, and also tell us about the diffusion of atoms across a surface: what path they take, where the blockages to diffusion occur and, in the case of a molecule, in what orientation they diffuse.

The method is unproven as yet, but experiments and the associated theory are in hand for a serious attempt to extract this elusive information.

5. Conclusions

Surface crystallography is providing structures which are relevant to work at the forefront of surface technology. We are not limited to simplistic surfaces of only academic interest, though obviously many complex surface structures still elude us. For the future a host of new ideas are being put forward. Those presented in this paper represent very largely my own viewpoint, but many researchers are actively seeking to widen the field and to make the study of surfaces as rich and diverse as that of the bulk.

References

Andersson, S. & Pendry, J. B. 1980 *J. Phys.* C **13**, 3547.

Barker, R. A., Estrup, P. J., Jona, F. & Marcus, P. M. 1978 *Solid St. Commun.* **25**, 375.

Heinz, K. & Mueller, K. 1982 In *Structural studies of surfaces*. Berlin: Springer.

MacLaren, J. M., Pendry, J. B., Rous, P. J., Saldin, D. K., Somorjai, G. A., Van Hove, M. A. & Vvedensky, D. D. 1987 *Surface crystallographic information service*. Dordrecht: Reidel.

Onuferko, J., Woodruff, D. P. & Holland, B. W. 1979 *Surf. Sci.* **87**, 357.

Pendry, J. B. 1974 *Low energy electron diffraction*. London: Academic.

Pendry, J. B. 1980 *J. Phys.* C **13**, 937.

Pendry, J. B. & Heinz, K. 1990 *Surf. Sci.* **230**, 137.

Rous, P. J. & Pendry, J. B. 1989*a* *Comp. Phys. Commun.* **54**, 137.

Rous, P. J. & Pendry, J. B. 1989*b* *Comp. Phys. Commun.* **54**, 157.

Rous, P. J., Pendry, J. B., Saldin, D. K., Heinz, K., Mueller, K. & Bickel, N. 1986 *Phys. Rev. Lett.* **57**, 2951.

Van Hove, M. A. & Tong, S. Y. 1979 *Surface crystallography by* LEED. Berlin: Springer.

Van Hove, M. A., Lin, R. F. & Somorjai, G. A. 1986 *J. Am. chem. Soc.* **108**, 2532.

Van Loenen, E. J., Frenken, J. W. M., Van der Veen, J. F. & Valeri, S. 1985 *Phys. Rev. Lett.* **54**, 827.

Zannazzi, E. & Jona, F. 1977 *Surf. Sci.* **62**, 61.

Discussion

R. Haydock (*Materials Science Institute, U.S.A.*). I am intrigued by Professor Pendry's method of finding a structure by linear methods applied to a distribution of structures. This seems a general way of solving nonlinear equations by linear methods. Does he know of other applications?

J. B. Pendry. I am not aware that this 'nonlinear inversion' procedure has been used before, but would imagine that it has wide application wherever there is a complex function whose global minimum has to be found.

L. M. Falivoc (*University of California, Berkeley, U.S.A.*). Together with Professor Pendry's reported advances in the theoretical interpretation of LEED there has been enormous progress in the experimental methods which aim at visualizing the position of atoms at surfaces. In particular the proposed electron holographic methods based on photoelectron diffraction seem to be on the threshold of a major breakthrough in 'real space' surface-structure determination.

J. B. PENDRY. Yes I agree. However, the correct interpretation of LEED or PE hologram depends on *future* development of theoretical methods as well as the direct methods discussed here.

J. D. C. MCCONNELL (*The University of Oxford, U.K.*). Does the absorbed benzene continue to resonate?

J. B. PENDRY. Probably not, as the distortion is considerable, several tenths of an ångström.

Electronic states in complicated materials: the recursion method

By Roger Haydock

Department of Physics and Materials Science Institute, University of Oregon, Eugene, Oregon 97403, U.S.A.

The determination of electronic states in complicated materials is difficult because of large numbers of inequivalent and nearly degenerate electronic orbitals. The only available approach is direct integration of the Schrödinger equation by path summation for which the recursion method gives a convergent expansion of the energy resolvent as a continued fraction whose parameters may be expressed as summations of groups of mutually avoiding paths. The inverse Fourier transforms of these continued fractions are matrix elements of the propagator and hence provide convergent discrete approximants for Feynman path integrals. Path counting for sequences of close packed layers is illustrated, and the application of the recursion method to the structural stability of transition metal Laves phases is reviewed briefly.

1. Introduction

The calculation of electronic states for materials with large numbers of inequivalent atoms requires a general solution of the Schrödinger equation. In such complicated materials, the coupling between electronic orbitals is usually larger than the differences in the energies of the orbitals so that perturbation theory for the electronic states does not converge, and because there are so many orbitals, a variational approach requires minimization with respect to a large number of parameters. In such circumstances the only approach is direct integration which, because the systems are two or three dimensional, requires path summation or path integration.

This paper describes the recursion method (Haydock 1980), a general solution to the Schrödinger equation, which has been used for the past 20 years to calculate densities of states and other electronic properties of materials with defects, disorder, or large unit cells. The idea behind this method is that instead of trying to calculate all the electronic states of the system, only those accessible from a given initial orbital are determined. The electronic hamiltonian is recursively transformed into a tridiagonal matrix by constructing a set of basis orbitals, of which the first is the initial orbital. This process of tridiagonalization can be viewed in a variety or ways: as direct integration of the Schrödinger equation in the non-degenerate subspace of states accessible from the initial orbital, as a variational solution of the Schrödinger equation in the subspaces spanned by powers of the electronic hamiltonian on the initial orbital, or even as a method for summing infinite sets of terms in perturbation theory.

This work returns to the origin of the recursion method in the counting of quantum mechanical paths and presents a way of expressing physical quantities in terms of a set of elementary determinants of paths. Friedel (1954) pointed out that the moments, integrals over powers of energy, of the density of states projected on an orbital, are the sums of products of the matrix elements of the hamiltonian along the paths which begin and end at that orbital. This idea was applied by Cyrot-Lackmann (1970) to the local densities of states of transition metals, and by Burdett (1986) to a variety of solid structures. Because the moment expansion of physical quantities does not converge Haydock *et al.* (1972*a*) developed the recursion method as a direct relation between the hamiltonian and the density of states. One objection to the recursion method has been that the intuitive appeal of path counting was lost and that the matrix elements of the hamiltonian in tridiagonal form have no simple physical interpretation. It is shown below that instead of these matrix elements, physical quantities can be expressed in terms of sums of mutually avoiding quantum mechanical paths. Some simple examples of chemical reactivity and structural stability are used to illustrate the ideas of path counting, and the application of the recursion method to the structural stability of the transition metal Laves phases is briefly reviewed.

The time development of a quantum mechanical system is given by the propagator, $\exp\{-i\boldsymbol{H}t\}$, applied to the initial state of the system ϕ_0, where $\boldsymbol{H}$ is the hamiltonian and t is time. However, for the purposes of this work, it is more convenient to work with the Fourier transform of the propagator which is the energy resolvent $1/(E-\boldsymbol{H})$. The result of multiplying ϕ_0 by the resolvent is singular on the invariant states of $\boldsymbol{H}$, contained in ϕ_0, with energy E. The time-dependent evolution of ϕ_0 can be recovered by an inverse Fourier transform, and similarly, time-dependent expressions can be recovered from all that follows.

2. Quantum path counting

Although either the full propagator or the full resolvent can be expanded in terms of quantum path summations, in the interests of simplicity, consider just the diagonal element of the resolvent,

$$R(E) = \phi_0^*(E-\boldsymbol{H})^{-1}\phi_0, \tag{2.1}$$

where in Dirac's notation ϕ_0^* is the bra corresponding to the ket ϕ_0. Physically this describes the energy spectrum of the invariant states contained in ϕ_0, and its Fourier transform is the propagator element for the system to start in ϕ_0 and return to it after a time t.

Expansion of this resolvent element in paths in equivalent to expansion of the resolvent in powers of the hamiltonian,

$$R(E) = \Sigma\,\phi_0^*\,\boldsymbol{H}^n\phi_0(1/E)^{n+1}, \tag{2.2}$$

where the sum is over n from nought to infinity. The expectation value $\phi_0^*\boldsymbol{H}^n\phi_0$ is converted to a path summation by introducing a countable, complete set of basis orbitals $\{\phi_m\}$ to make,

$$R(E) = \Sigma\,\mu_n/E^{n+1}, \tag{2.3}$$

with

$$\mu_n = \Sigma\,H_{0,1}H_{1,2}\ldots H_{n-1,0}, \tag{2.4}$$

where the sum is over all paths of n steps beginning with ϕ_0, going from one basis orbital to another, and ending with ϕ_0; and each term in the sum is the product of the matrix elements of $\boldsymbol{H}$ for each step of that path. The expansion in quantum paths is always possible because there is always a countable complete set of orbitals, for example the isotropic oscillator wavefunctions, or the orbitals localized within each atomic cell. Since each path has a finite number of steps, the product of the matrix elements of the hamiltonian is finite for each path; however, the expansion of the resolvent in inverse powers of E, and the corresponding expansion of the propagator in powers of t are asymptotic rather than convergent. The sums over paths of length n, μ_n, are also the moments of the projected density of states, as can be seen by integrating E^n over $R(E)$ on a contour which encloses the real energy axis of the complex energy plane.

The above expansion of the resolvent as a sum over paths does not converge because the number of paths increases faster than any power of their length. This can be seen in a simple system consisting of two identical orbitals A and B. At each step of a path the choice is to step to the same orbital or to the other orbital. If the step is to the same orbital, then that step contributes a factor of the energy of the orbital v which is the same for both A and B, and if the step is to the other orbital, then the contribution is h, independent of whether the step is from A to B or from B to A. By convention, there is one path of length zero, contributing one to μ_0; there is also one of length one with μ_1 equal to v; two of length two with μ_2 equal to v^2+h^2; and so on. In general, there are 2^{n-1} paths of length n with μ_n equal to $\frac{1}{2}(v+h)^n+\frac{1}{2}(v-h)^n$. As n increases, both the number and contribution of the paths increases exponentially, and since these contributions are divided by just E^n, the path expansion in this example converges only for E sufficiently large that it is uninteresting. For other examples, the series converges nowhere.

In applications of quantum path counting, it is usual to truncate the set of basis orbitals as in the tight-binding approximation where only atomic-like valence orbitals are retained. For example in comparing the electronic states at transition metal surfaces (Haydock *et al.* 1972*b*) needed only the atomic d-orbitals to get a good qualitative picture of surface electronic structure. Atoms on different surfaces of the same transition metal have different coordinations, for example the 111 surface of FCC nickel has coordination 9, while the 100 surface has coordination 8. The lowest moment of path length for which these two surfaces differ is μ_2 which is greater for the 111 surface by the ratio of 9 to 8 relative to the 100 surface. Thus the ratio of the spread in energy of the electronic states at these two surfaces is $(\frac{9}{8})^{\frac{1}{2}}$ and, crudely the ratio of the density of states at the Fermi level is $(\frac{8}{9})^{\frac{1}{2}}$. The reactivity of a metal surface is roughly related to the amplitudes of states at the Fermi level ready to hybridize with the electronic states of a passing atom; so from this simple path counting argument and Fermi's golden rule, one concludes that some sort of zero temperature reaction rate on the two surfaces should be in the ratio of 8 to 9, the ratio of squares of the hybridizing matrix elements.

3. Determinants of paths

An alternative to a series is the continued fraction expansion which from the time of the ancient Greeks has been recognized as a highly convergent approximation scheme. In this approach, the paths are generated from a set of elementary paths, as for example, the paths in above the two orbital system are made up entirely of hops which stay on the same atom and hops which change atoms. This path expansion may be summed to give $1/(E-v-h^2/E)$ which is the first two levels of a continued fraction. If one attempts to extend this idea to a three-orbital system, expressions become much more complicated and a general analysis is necessary.

In the nineteenth century, work on the above moment problem revealed that the elementary quantities from which all the paths could be generated are a set of determinants (Shohat & Tamarkin 1943) which may be expressed in terms of the moments as

$$A_n = \begin{vmatrix} \mu_0 & \mu_1 & \cdots & \mu_{n-1} & \mu_{n+1} \\ \mu_1 & \mu_2 & \cdots & \mu_n & \mu_{n+2} \\ \cdot & \cdot & \cdots & \cdot & \cdot \\ \cdot & \cdot & \cdots & \cdot & \cdot \\ \mu_n & \mu_{n+1} & \cdots & \mu_{2n-1} & \mu_{2n+1} \end{vmatrix}, \tag{3.1}$$

$$B_n = \begin{vmatrix} \mu_0 & \mu_1 & \cdots & \mu_{n-1} & \mu_n \\ \mu_1 & \mu_2 & \cdots & \mu_n & \mu_{n+1} \\ \cdot & \cdot & \cdots & \cdot & \cdot \\ \cdot & \cdot & \cdots & \cdot & \cdot \\ \mu_n & \mu_{n+1} & \cdots & \mu_{2n-1} & \mu_{2n} \end{vmatrix}. \tag{3.2}$$

These determinants consist of sums of products of the moments which in turn are sums of products of the matrix elements of the hamiltonian along paths. Because the same paths occur in many moments in each determinant, there is a great deal of cancellation which makes direct evaluation of the determinants impractical.

The recursion method (Haydock 1980) avoids these determinants by direct tridiagonalization of the hamiltonian. The process begins with a normalized orbital ϕ_0, and generates a new orbital,

$$\phi_1 = (\boldsymbol{H}\phi_0 - a_0\phi_0)/b_1, \tag{3.3}$$

where a_0 and b_1 are chosen to orthonormalize ϕ_1 with respect to ϕ_0. The nth recurrence is,

$$\phi_{n+1} = [(\boldsymbol{H} - a_n)\phi_0 - b_n\phi_{n-1}]/b_{n+1}, \tag{3.4}$$

where a_n and b_{n+1} are chosen to orthonormalize ϕ_{n+1} with respect to ϕ_n. A convergent expansion of the resolvent is given by the continued fraction,

$$R(E) = 1/[E - a_0 - b_1^2/(E - a_1 - b_2{}^2/(E - a_2 - \dots))], \tag{3.5}$$

from which it follows that these tridiagonal matrix elements generate the moments of the hamiltonian. While this method of constructing the continued fraction is numerically stable, it does not involve any explicit path summation, and so does not give any physical insight into the importance of various physical processes. It is difficult to interpret the parameters other than as matrix elements of an equivalent one-dimensional hamiltonian.

The cancellation between paths in different terms of the above determinants can be made explicit by introducing an artificial time dependence, and by considering groups of paths. Ratios of the determinants are then the parameters of the continued fraction expansion of the resolvent. Think of paths in terms of a step at every tick of a clock. Each step goes from the current orbital to any other orbital including the same one, and multiplies the contribution from the path by the matrix element of the hamiltonian between the two orbitals of the step. The determinant B_0 is defined to be the normalization of ϕ_0, and the determinant A_1 is simply the diagonal matrix element of $\boldsymbol{H}$ for ϕ_0, which we can think of as the contribution from paths which start at ϕ_0 at $t=0$ and end at ϕ_0 at $t=1$. The next determinant is B_1, which we may think of as the contribution from pairs of paths which start at $t=-1$ and $t=0$, finishing at $t=0$ and $t=1$, where we now take for granted that paths start and finish on ϕ_0. There are two ways that paths can satisfy this timing: the first is that one path leaves the origin at $t=-1$ and returns at $t=1$ after two steps while the other path starts and finishes at $t=0$ with no steps, and the second way is that both paths consist of one step starting at the origin at $t=-1$ and $t=0$. From the definition of B_1, the first way of starting and finishing the paths contributes with a positive sign while the second is negative because of the permutation of the times of finishing relative to the times of starting. Paths which meet at the same orbital at the same time, contribute to the determinant in two ways, but with opposite sign because of the permutation, and so cancel. For example, in B_1 the pair of paths which stay at the origin from $t=-1$ and $t=0$, finishing at $t=0$ and $t=1$, meet at the origin and contribute with opposite signs depending on the order of finishing.

The $(n+1)$ by $(n+1)$ determinants of B_n or A_n may be expressed as the sum of contributions from groups of $n+1$ paths (Viennot 1989) starting at $t=-n$, $t=-n+1$, ..., $t=0$, and finishing at $t=0$, $t=1$, ..., $t=n$ for B_n or $t=0$, $t=1$, ..., $t=n-1$, not $t=n$, but rather $t=n+1$, for A_n. The contribution from one of these groups of $n+1$ paths is the product of the matrix elements for their steps with a sign given by the permutation of the order of finishing relative to first out, last in. Groups of paths which meet on the same orbital at the same time cancel because each meeting allows the group to contribute twice but with opposite sign. Only the groups of paths which never meet actually contribute to the determinant. It is worth noting that because the determinants are the sums of contributions from groups of paths, a perturbation to the hamiltonian produces only a polynomial rather than an infinite series.

In terms of the determinants of mutually avoiding paths, the resolvent is given by the continued fraction equation (3.5) where the parameters are,

$$b_n = [B_{n-2}B_n]^{\frac{1}{2}}/B_{n-1}, \tag{3.6}$$

and

$$a_n = [A_n/B_n]-[A_{n-1}/B_{n-1}]. \tag{3.7}$$

Inverse Fourier transforms of the above continued fractions provide convergent approximants to the Feynman path integrals (Feynman & Hibbs 1965). For example, the matrix element of the propagator from ϕ_0 to ϕ_0 in time t is,

$$\int \exp\{-\mathrm{i}A(P)\}\,\mathrm{d}P = (2\pi\mathrm{i})^{-1}\int \exp(-\mathrm{i}Et)\,R(E)\,\mathrm{d}E, \tag{3.8}$$

where the integral on the left is over all paths P from ϕ_0 at time 0 to ϕ_0 at time t, with the action $A(P)$ appropriate for $\boldsymbol{H}$; and the integral on the right is around a contour in the complex E-plane which encloses the singularities of $R(E)$. The path summations can be viewed as a discretization of the path integral.

4. Transition metal Laves phases and the stability of stacking sequences

As a simple example, consider the bandstructure contribution to the total energy of the FCC structure from the partly filled s-orbitals of metal atoms. This structure consists of spherical atoms in close-packed layers which can stack on top of one another in three equivalent ways. For the ideal HCP structure the layers alternate in

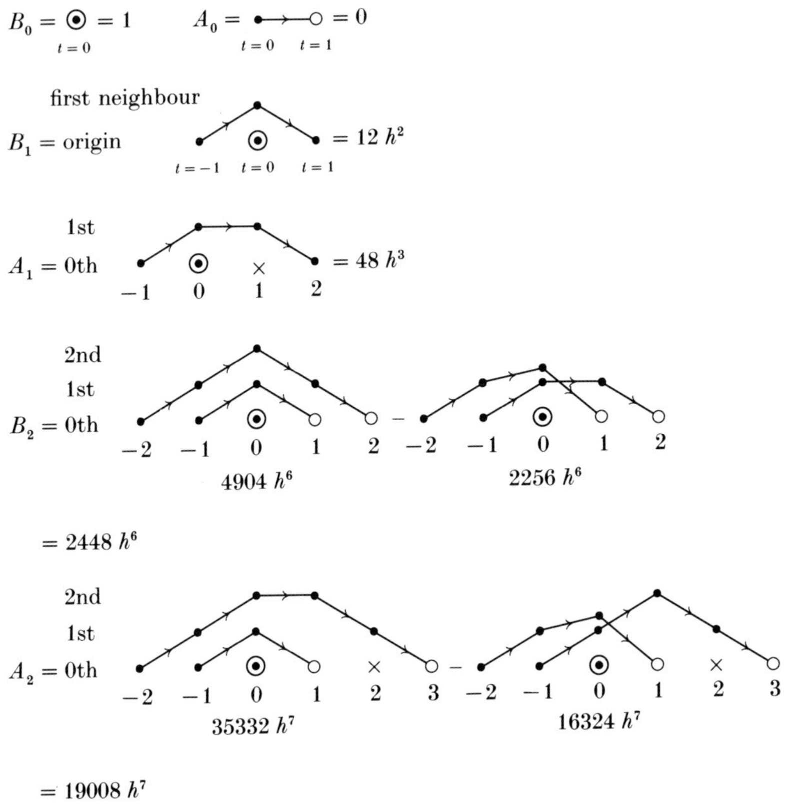

Figure 1. Schematic representations and numbers of the mutually avoiding paths which contribute to the FCC band-structure energy.

position, while for the FCC structure every third layer repeats. The band-structure energy can be calculated from the projected density of states which for the s-orbitals in these structures is also the average density of states because the s-orbitals are all equivalent. The sum of the energies of the occupied states is then the band-structure contribution to the cohesive energy of each structure.

The continued fraction for the projected resolvent and density of states can be calculated using the above method of path counting with a simple hamiltonian which has matrix elements h between nearest neighbour orbitals and zero otherwise. Each orbital has 12 nearest neighbours and 42 second neighbours for the FCC structure. Some of the various kinds and numbers of mutually avoiding paths are shown in figure 1.

Using the relations between the determinants and the continued fraction parameters, equations (3.6) and (3.7), gives an approximate resolvent element,

$$R(E) = 1/[E-12h^2/(E-4h-17h^2/(E-768h/204))]. \tag{3.9}$$

The band-structure energy per atom is then the sum of the residues of this resolvent for the poles corresponding to occupied states, weighted by the energy of each pole.

To investigate the role of d-band electron per atom ratio, the recursion method has been applied to the structural stability of the AB_2 transition metal compounds which form in one of the three Laves phases (Haydock & Johannes 1975; Johannes *et al.* 1976) which are isomorphs of the $MgCu_2$, $MgZn_2$, and $MgNi_2$ structures. Like HCP and

FCC, these compounds may be described by their stacking sequences which a much more complex and repeat every 12th, 8th, and 16th layer respectively. The band-structure contribution to the cohesive energy was calculated as a function of d-band filling for each structure using a hamiltonian derived from canonical d-orbitals scaled to the various atoms.

In this work the continued fraction parameters described above were obtained from recurrences rather than path counting, so it was not possible to identify the electronic processes which account for the structural energy differences. However, the results did demonstrate the importance of the relationship between the local environment of atoms and the symmetry of the atomic orbitals in that the band-structure energy. The differences in band-structure contribution to the cohesive energy calculated using a difference in energy between d-orbitals on the two kinds of atoms which depends quadratically on the group number difference of the atoms, correctly separates the structures of all the 76 known non-magnetic occurrences of such compounds.

5. Conclusion

In the absence of symmetry, the quantum mechanical states of a system can only be determined by direct integration of its equations of motion. Although the power series expansion of the resolvent or propagator in paths does not converge, the continued fraction expansion does, and its parameters can be expressed as ratios of sums of mutually avoiding groups of paths, through which the behaviour of the system can be related to elementary physical processes. Subtle differences in structural energies can be calculated accurately by this means.

I thank the Cavendish Laboratory and Pembroke College of Cambridge University for their hospitality during the writing of this paper, Volker Heine and W. Matthew C. Foulkes for the questions which motivated this work, and the United States National Science Foundation Condensed Matter Theory Program for support under grant DMR-8712346.

References

Burdett, J. K. 1986 *Molecular structures and energetics*, vol. 1 (ed. J. F. Liebman & A. Greenberg), p. 209. Verlag Chemie.

Cyrot-Lackmann, F. 1970 *J. Phys. (Paris) Suppl.* C**1**, 67.

Feynman, R. P. & Hibbs, R. A. 1965 *Quantum mechanics and path integrals*. New York: McGraw Hill.

Friedel, J. 1954 *Adv. Phys.* **3**, 446.

Haydock, R. 1980 *Solid state physics* **35** (ed. H. Ehrenreich, F. Seitz & D. Turnbull), p. 215. New York: Academic Press.

Haydock, R., Heine, V. & Kelly, M. J. 1972*a* *J. Phys.* C**5**, 2845.

Haydock, R., Heine, V., Kelly, M. J. & Pendry, J. B. 1972*b* *Phys. Rev. Lett* **29**, 868.

Haydock, R. & Johannes, R. L. 1975 *J. Phys.* F**5**, 2055.

Johannes, R. L., Haydock, R. & Heine, V. 1976 *Phys. Rev. Lett.* **36**, 372.

Shohat, J. A. & Tamarkin, J. D. 1943 *The problem of moments*. Mathematical Surveys **1**. Providence: American Mathematical Society.

Viennot, X. G. 1989 Orthogonal polynomials: theory and practice. Lecture at NATO Advanced Study Institute, Columbus, Ohio, May–June 1989.

Discussion

N. W. Ashcroft (*Cornell University, U.S.A.*). It is indeed impressive that this reformulation of the recursion method accounts for the subtle energy differences found in systems with quite complex unit cells. Per atom, it would seem that these energy differences are on the scale of a few tens of degrees. Can a simple physical argument be offered that explains why ion dynamics (on the scale of a few hundred degrees in a given structure) is apparently playing so insignificant a role?

R. Haydock. The stable structure is that with the minimum Gibbs free energy, to which the band-structure energy is one contribution. In the case of the transition metal Laves phases, I have argued that the electronic band-structure energy is the dominant contribution giving differences of order 10^{-3} rydberg per formula unit which corresponds to temperatures of hundreds of kelvins at which some of the compounds studied do indeed undergo structural transitions.

As to other contributions to the Gibbs free energy such as the ion dynamics, what matters are again the differences between structures. The phonon spectra could be calculated for the transition metal Laves phases by similar methods to those I have used for the electrons. Their contributions to the free energy is like that of the electrons but scaled by about 10^{-3}, the ratio of electronic to ionic masses. Thus, although the electronic energy differences are of similar size to phonon energies, it is the differences in phonon energies which matter and they are about one thousandth of the differences in electronic energies.